Stockamsel

Stockamsel

by Claudio Guerrieri

To my wonderful mother Norma

CONTENTS

CHAPTER ONE

Stockamsel

Birds that inhabit the pages of ornithological books do not always fare well. Some end up in obscurity, not so much because they become extinct, but because they may have never really existed at all, at least not as a distinct species. This is the story of one of these birds, the *Stockamsel*, a type of Blackbird or Ouzel that inhabited the forests of Germany. Its presence would also be noted in other European countries such as Great Britain and Ireland where, due to its peculiar plumage, it would catch the eye of the keen observer. Elsewhere in Europe, other birds with fanciful names such as *Bergamsel*, *Merle brun* and *Graue amsel* were described and may well have corresponded to the *Stockamsel*. These, too, inhabited central Europe, though in reality they resided mainly amongst the pages of the ornithological literature of the time.

And then the *Stockamsel* quietly disappeared, rarely mentioned in ornithology books any more, and buried within long lists of forgotten bird synonyms. And yet, in the 1700s it claimed much notoriety, even to the point of being associated with birds described by the great luminaries of Natural History and the burgeoning field of Ornithology such as Carl Linnaeus in Sweden, Conrad Gesner in Switzerland, and John Ray and Francis Willughby in Britain.

But what happened?

It was while researching the story of the *Ring Ouzel* that this author's interest was stimulated by the encounter with a cryptic bird called the *Stockamsel*, a curiosity that was enhanced by the almost total lack of information about this particular bird, with only a smattering of citations in current ornithology books.

One very recent mention was found in the book *The Helm Guide to Bird Identification: An In-Depth Look at Confusion Species* (2014) by **Vinicombe** et al.[1] When discussing aberrant types of the *Common blackbird* (*Turdus merula*) and how they compare to the *Ring ouzel*, the authors say:

> Aberrant Blackbirds occasionally show white feathering in their plumage and, when this is present on the breast, it can cause confusion with Ring Ouzel. Confusion could also arise with first-winter male Blackbirds of the

so-called **'stockamsel'** type (from Germany and Poland) which have a dull bill and eye-ring, browner wings, a paler chin and heavy pale fringing to the underparts feathers.

This passage begs the following questions:

(1) What is a "first-winter male Blackbird"?

(2) What is a 'stockamsel' type of Blackbird?

(3) What do they mean by "from Germany and Poland"?

To begin with, the *Common blackbird* (*Turdus merula*) is a type of thrush that inhabits Europe and Asia, and is thus known as the *Eurasian blackbird*, a term which distinguishes it from the New World blackbirds. For the purpose of simplicity, the *Common blackbird* will be referred to from now on as simply *Blackbird* within the text of this book.

To answer the first of our questions, two areas need to be reviewed: (1) the normal plumage of the *Blackbird*, and (2) the changes in its plumage with moulting.

The *Blackbird*, as described by British naturalist David W. **Snow** in *A Study of Blackbirds* (1988), shows the following plumage features:

1) The adult male has a sooty black plumage without the metallic reflections seen in crows. Its plumage has a satiny gloss and may exhibit a scaly appearance due to the feather edges being somewhat less glossy than the inner parts.

2) The adult females, on the other hand, are earth-brown and have pale throats and a vaguely spotted breast.

3) The young or juvenile *Blackbirds* are redder brown than the females, with a more accentuated spotting below. The feathers show pale shaft-streaks on the mantle and spots on the wing coverts. [2]

A portrayal of the male and female adult as well as a juvenile *Blackbird* that exemplifies such a description is depicted on the following page.

The term "first-winter male" refers to a phase of the *Blackbird's* life that occurs between the juvenile and the adult, during which the bird carries a specific plumage. *Blackbirds* undergo several changes in their plumage after birth. For many European passerines, including the *Blackbird*, the most common and basic moult strategy is a sequence of plumages that goes from *Nestling - Juvenile - First Winter - Adult*. The standard cycle of

Turdus merula L. Schwarzdrossel.
1 altes Männchen. 2 altes Weibchen. 3 junger Vogel.
¼ natürl. Grösse.

Figure from Naumann's *Naturgeschichte der Vögel Mitteleuropas* (1905) entitled "Turdus Merula (Schwarzdrossel)." *top*: juvenile; *middle*: adult male; *bottom*: adult female. [3]

4

plumages of the *Blackbird* is basically a continuous series of different plumages separated by their respective moulting.

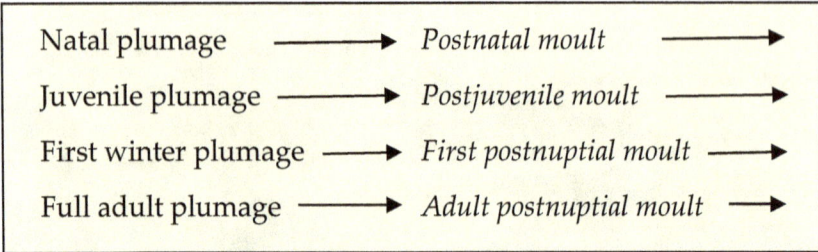

Natal plumage ⟶	*Postnatal moult*	⟶
Juvenile plumage ⟶	*Postjuvenile moult*	⟶
First winter plumage ⟶	*First postnuptial moult*	⟶
Full adult plumage ⟶	*Adult postnuptial moult*	⟶

At birth, the *Blackbird* is covered by sparse down-feathers (natal plumage). One week after hatching, the bird begins to sprout the juvenile plumage (post-natal moult) which will be worn for about three months. In fact, after fledging (i.e., the moment a feathered chick leaves the nest) they wear this juvenile plumage all summer. Thereafter, the juveniles begin to moult into their first adult-type plumage, which is worn throughout the first year of the bird's adult life, that is, until the following autumn. This post-juvenile moult usually occurs in August or September of their first autumn of life and lasts about five weeks. During this moult, *Blackbirds* undergo only a partial moult; i.e., the body feathers and some wing feathers (lesser and median coverts, underwing coverts, and a variable number of greater and secondary coverts) are replaced; but not the primaries, secondaries or tail feathers.[4] Active moult ceases in the autumn (a time that corresponds to the start of the autumn migration of non-stationary birds). These immature birds thus exhibit a 'moult contrast,' that is, the presence of juvenile feathers alongside adult feathers, and this is most noticeable on its wings. However, the new adult-type feathers are usually duller or paler than those of the full adult plumage, and those of the breast are edged with brown.[2] In the field, male *Blackbirds* with this first-winter plumage can be recognized by the paler and more brownish wings which contrast with the black of the rest of its plumage.

David **Snow**, in this regard, makes the following significant observation in his book *A Study of Blackbirds* (1988):

> First-year plumages vary considerably, some being rather close to the adult plumage and some very distinct. Yearling males, in particular, may be almost as black and glossy as full adults, or almost as brown as old females, with the whitish throat of a female. German ornithologists, who have paid some attention to these variations, have called the first-year plumage which

is close to the full adult plumage 'Fortschrittskleid' (advanced plumage) and the other extreme 'Hemmungskleid' (retarded plumage). Males in retarded first-year plumage are known to German bird-catchers as 'Stockamsel'. [2]

The author is referring to the fact that a first-year bird which retains juvenile-type feathers is said to be in "retarded" plumage. To explain this so-called "retarded plumage," several physiological hypotheses have been proposed. As mentioned earlier, the post-juvenile moult occurs in August to September of the *Blackbird*'s first autumn of life regardless of when it was born, which would have been any time between March and July. It has been suggested that later hatched juveniles have less time to moult between fledging and the necessity of arresting moult before the onset of winter.[4] Such late-hatched *Blackbirds* would correspond to those with "retarded plumage." One proposed explanation for this is that the gonads of late-hatched birds are less developed, with a consequent smaller production of sex hormone which is known to control the formation of male or female plumage. On the other hand, such plumage changes may occur in the face of a normal hormonal milieu, and this would probably be due to altered hormonal receptors on the target feathers.

Another explanation for retarded plumage could implicate the degree of "moult extent." The moulting of *Blackbirds* in their first year is vulnerable if food supplies are short and there is insufficient nourishment to support the moult. Since moulting is a highly energy-demanding process, nutritional stress can lead to moult interruption. It has been shown that when the environmental situation sharply deteriorates such as in the Northern regions of Europe, some *Blackbirds* may have little time to moult and will migrate from their breeding territories, often with an unfinished or partial moult. Moult extent shows a clear tendency to be reduced towards the end of the season and this reduction is greater in northern populations of *Blackbirds*.[5] In fact, in their first autumn young blackbirds seem to renew fewer wing feathers in Sweden than in Switzerland.[6] On the other hand, birds in advanced or "progressive" plumage have been more often noted in geographical areas with a warmer climate where there is more food and better conditions for moulting.

If the extent of post-juvenile moult is related to the fledging dates of birds, geographic and migratory factors may come into play to determine why the *Stockamsel* variety is seen more often in certain regions than in others. For instance, breeding of *Blackbirds* in Scotland takes place later than in the south of England and so, in general, Scottish *Blackbirds* progress less far with their post-juvenile moult than those from southern England.[3] It follows that the English resident *Blackbirds* may vary from their Scottish migratory cousins due to their different hatching pattern.

As to the question of what constitutes a first-winter male *Blackbird*, several authors have eloquently described this. For instance, in 1902, Henry **Seebohm** wrote *A Monograph of the Turdidae*[7] in which he described the first-winter young male *Blackbird*:

> In the first winter the young males display great diversity of plumage, and scarcely two individuals are alike, the feathers of the under surface being edged with rufous-brown, which fades to buff, or even to dull white. The throat is often white, streaked with black, and the breast-feathers sometimes show whitish shaft-lines. The bill is sometimes yellow, sometimes black, and often parti-coloured.

According to Harry Forbes **Witherby** in his book *A Practical Handbook of British Birds* (1920), a first-winter male *Blackbird* would be:

> Like adult male but browner, not so jet-black, some feathers of crown usually, and often some of mantle, with inconspicuous brownish edgings and most feathers of under-parts with more marked brownish or greyish edgings and often with pale shaft-streaks, chin and upper-throat sometimes with conspicuous grey edgings and occasionally grey with dark streaks almost as in female; wing-feathers and primary coverts browner than in adult male and unmoulted outer greater coverts conspicuously browner (and with pale tips) than new and black inner ones. The juvenile body-feathers, lesser and median wing-coverts and varying number of inner greater coverts are moulted Aug.-Oct. (occasionally extending to Nov.-Dec.), but not tail, wing-feathers or outer greater coverts.[8]

Another feature of first-winter males is the dark color of the beak and the eye-rim. Only during the first winter will the eye-rim and the beak, usually starting at its base, begin to change to yellow, but never as bright as that of the adult.

Even though it is not very clear what distinguishes a 'non-stockamsel' type of first-winter male from a 'stockamsel' type, the main feature that appears to be stressed is the degree of pale feather edging of the underparts.[1]

After the winter, the moult-plumage cycle continues with some first-winter *Blackbirds* acquiring a first-summer or nuptial plumage through abrasion and fading of its feathers. Then, starting in July of its second year, the first-summer bird will be subject to a complete moult which lasts about two months during their second autumn. Thereafter, it will be dressed in the final adult plumage (or 'adult winter plumage'). These adult birds will then undergo a complete moult following the breeding season from August to November every year.

Below is a photograph of a first-winter male *Blackbird* with features, especially the "heavy pale fringing to the underparts feathers," that would categorize it as a *Stockamsel* variety.

A first-winter male *Blackbird* photographed in Ireland in late January or February. Note the dull yellow eye ring, variegated bill, several brown juvenile wing feathers, and a pronounced pale edging of the underparts feathers. Photograph © by Anthony McGeehan.

In the next page is a schematic diagram of the moults and plumage changes of the *Blackbird* during its first year of life. For addional information, see appendix B (page 98).

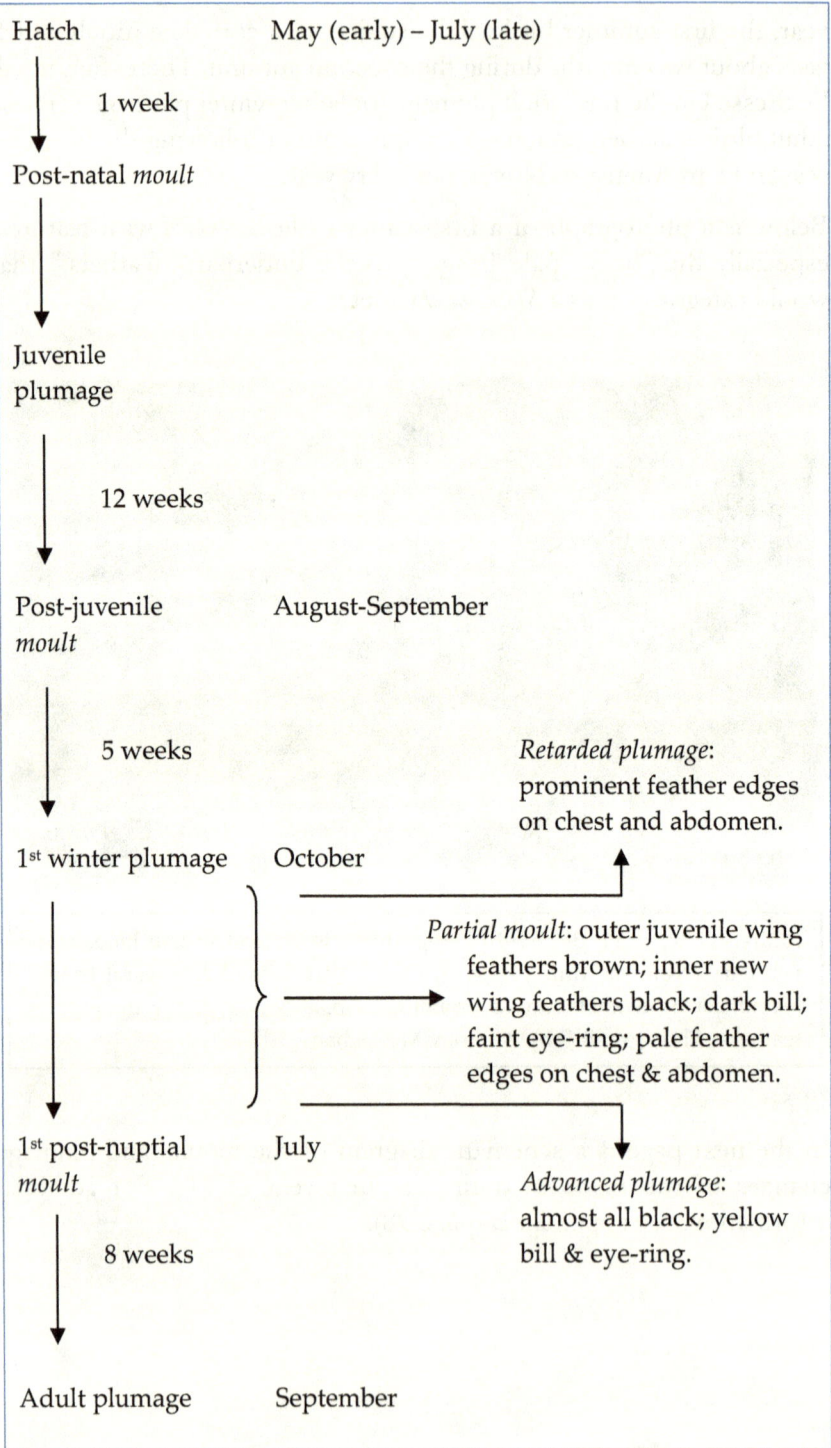

Hatch   May (early) – July (late)

↓   1 week

Post-natal *moult*

↓

Juvenile plumage

↓   12 weeks

Post-juvenile *moult*   August-September

↓   5 weeks

Retarded plumage: prominent feather edges on chest and abdomen. ↑

1st winter plumage   October

→ *Partial moult*: outer juvenile wing feathers brown; inner new wing feathers black; dark bill; faint eye-ring; pale feather edges on chest & abdomen.

1st post-nuptial *moult*   July

↓ *Advanced plumage*: almost all black; yellow bill & eye-ring.

↓   8 weeks

Adult plumage   September

CHAPTER TWO

"Stock-"

To answer our queries regarding the meaning of the term *Stockamsel* and its supposed provenance "from Germany and Poland," one needs to review the ornithological literature and see if we can trace the origin of the word "Stockamsel."

The definition of the prefix "Stock" can be found in German books of the time, especially those with hunting-related terminologies since it appears that it was the local hunters in German-speaking regions who came up with the name *Stockamsel*. One of the earliest mentions of this prefix was by **Großkopff** in 1759 when he wrote the hunting and forestry dictionary called *Neues und Wohl eingerichtetes Forst- Jagd- und Wiedewercks Lexicon.*[9] The term "Stock" was defined as:

> *Stock* or *Stöcke*: where the timber breaks or is cut low; the remnant that remains standing is called a stump or trunk.

Other sources would recapitulate this definition of the term "Stock" by stating that it represented, for instance, "the piece of wood of a chopped-down tree that is left in the ground." Likewise, in 1797, the ornithologist **Bechstein**, in his *Handbuch für Praktische Forst- und Jagdkunde* [Handbook of Practical Forestry and Hunting Knowledge], defined the term Stock as "the end of the trunk of a fallen tree, together with its roots, which remains in the soil."[10] And in 1892, **Dombrowski** defined it as "rootstock, the lowest part of the trunk, mostly stuck into the ground, and the roots it sprouted. It grows shoots, and these are the shoots growing out of a tree stump."[11]

In chapter 3, we will see that the earliest reference found in the German literature of the *Stockamsel* was preceded by that of a similarly termed bird called *Stockziemer*. The obvious question is whether the latter is related or even identical to the *Stockamsel*, considering that their suffixes [*Amsel = Ouzel*, and *Ziemer = Thrush*] are those of two closely related types of birds, the names of which have sometimes been used interchangeably.

Before the appearance of the bird names *Stockamsel* and *Stockziemer*, the German ornithology literature had cited the word "Stock" in reference to other birds, and specifically in regard to the site of their nest. For instance, in 1742 Johann Heinrich **Zorn** wrote the two-volume book titled *Petino-Theologie* in which he focused his attention on the behavior of birds. When talking about a bird called *Berg-Amsel* ("Mountain ouzel," the description of which matches perfectly that of the *Ring ouzel*, thus making these basically synonymous) Zorn also mentioned the *Wald-amsel* ("Forest ouzel"):

> Our *Waldamsel* [forest ouzel] indeed also makes its nest commonly in low bushes; but she makes the same on old stumps ["Stöcke"], or even in the young shoots and sprouts of pruned trees; . . . [12]

Zorn is referring to nests built on old tree stumps or even in the midst of a coppice with its densely packed tree shoots. We thus have a first indication of a bird that makes its nest on the stump of a fallen tree, either amongst its roots, or upon the base of the stump, or even in the thick of a coppice.

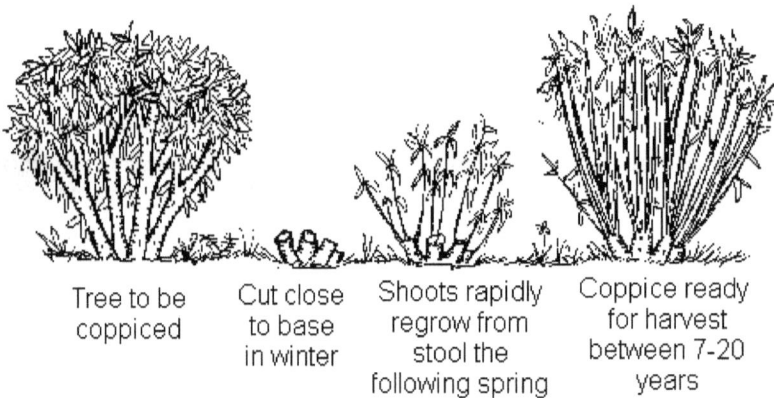

| Tree to be coppiced | Cut close to base in winter | Shoots rapidly regrow from stool the following spring | Coppice ready for harvest between 7-20 years |

The term *Waldamsel* has been a synonym for the *Ring ouzel* since the times of **Gesner** who, in this regard, mentioned it in his *Historiae Animalium* from 1555:

> It is also found in our [Swiss] mountains; and the local people call it the *Ringamsel* ["Ring ouzel"], i.e., *Merula torquata*, or the *Waldamsel*, i.e., *Merula sylvatica*. [13]

On the other hand, the *Waldamsel* has also been equated with the *Blackbird*. For instance, in the dictionary *Onomatologia Forestalis-Piscatorio-Venatoria* from 1772, regarding the *Blackbird* the authors say that when it resides in the forest, it is called a *Waldamsel* [forest ouzel].[14]

Another bird whose name also contains the prefix "Stock" is the *Stock dove* (*Columba oenas*). **Lockwood** in *The Oxford Dictionary of British Bird Names* (1984) says of the *Stock Dove*:

> Known since c.1340 'Stock-dowe'. *Stock* here means tree-trunk, alluding to the nesting place, a hole in a tree-trunk, which distinguishes this bird from the allied species; also called Hole Dove. [15]

The *Catholicon Anglicum* wordbook from 1483 gives "Stok" or "Stoke" as meaning *stipes* [stump] and *truncus* [tree-trunk] in Latin.[16] Therefore, "Stock dove" means roughly "a dove which lives in hollow trees." In fact, their nest is usually in a hole of an old tree, and thus it was normally found only in old forests.

Compare these terms with the German *Stocktaube* and *Holztaube*. Only in 1760 did **Klein** mention the German name of *Stocktaube* in his *Verbesserte und vollständigere Historie der Vögel* and equated it with the *Columba livia*, also known as Stock Dove or Wood-Pigeon.[17]

An additional German bird name that used the prefix "Stock-" was the *Stock-Eule* [Little Owl or Tawny Owl], as written in **Klein**'s *Historia Avium Prodromus* in 1750.[18] This name can be traced as far back as 1505 since it was mentioned in the German-Latin dictionary *Vocabularius Predicantium* by **Melber** & **Eichmann**.[19] According to a variety of sources, the *Stockeule* would correspond to either the Wood owl, Forest owl, Brown owl (*Syrnium aluco*), Tawny owl (*Strix stridula*), Hooting tawny owl (*Ulula aluco*) or the Little owl (*Athene noctua*).

The eagle called *Stockaar, Stockarn* or *Stockadler* is yet another German bird name with the prefix "Stock-". This, too, can be traced far back in time, and one of its earliest mentions is in the *Thierbuch* published in 1545:

> Our third type of eagle likes to rest on the truncated tree / therefore it is called by us a *Stock-adler* [Trunk Eagle]. [20]

Today, the *Stockaar* or *Stockadler* (also called *Stockfalke*) corresponds to the Goshawk (*Astur palumbarius, Falco palumbarius* or *Accipiter gentilis*).

It is thus of no surprise that the hunters and locals of German-speaking countries would eventually name a form of *Blackbird* that lived in the woods and nested on tree stumps with the designation of *Stockamsel*. Furthermore, since the ouzel (whether *Ring ouzel* or *Blackbird*) that nested on tree stumps was called a *Waldamsel*, as described by Zorn, it could be hypothesized that the *Waldamsel* is the actual forerunner of the *Stockamsel.* [12]

CHAPTER THREE

The Stockziemer of Döbel

It was in 1746 that Heinrich Wilhelm **Döbel** wrote the hunter's manual *Jäger-Practica*.[21] One chapter was dedicated to the *Ring ouzel*, even though it was titled under its alternate names *Meer-amsel*, *Schild-amsel* and, most interestingly, *Stock-ziemer*. Döbel's description of the *Stockziemer* translates as follows:

> This bird is as large and of equal flavor as the *Fieldfare*, admirably plump, appears black sprinkled with some indistinct white, has a white patch or shield on the chest and neck, does not breed in this country, but only travels through here. In the early part of the year you will not be aware of it much; but in the autumn it gets trapped like a foolish bird in snares . . . But its migration does not last long in the autumn; it also arrives late, and stays not more than 14 days or 3 weeks.

This description ostensibly corresponds to that of the *Ring ouzel*. Unfortunately, Döbel offers no explanation as to the etymological meaning or origin of the term *Stock-ziemer*. The *Meeramsel* (literally, "Sea ouzel," because it migrates with thrushes across the sea) and the *Schildamsel* (literally, "Shield ouzel," a blackbird with a white patch on its breast resembling a shield) were well known synonyms for the *Ring ouzel*. The confirmation that these three names corresponded to the actual *Ring ouzel* would be added later in future editions of this book. In fact, the index of the 5th edition included the entry "Stockziemer: see Ringdrossel."[22] Also, a note was added in the 1912 reprint of this book stating that these names corresponded to the "Ringamsel, Merula torquata."[23]

This is the first time that the term *Stockziemer* is used to designate a type of bird. The word is a combination of the two German words *stock* (tree stump) and *ziemer* (thrush). For unspecified reasons, the *Stockziemer* would later be equated with the *Stockamsel*, especially by those who believed that these were synonyms for the *Ring ouzel*. And yet, certain books such as the *Allgemeines Haus-haltungs Lexicon* published in 1750 completely separated the *Stockziemer* from the *Ring ouzel* as well as from the *Blackbird*. In fact, it called *Ring ouzel* a type of blackbird with a white ring around the neck, while the *Stockziemer* was vaguely described as black with some disseminated white, thus setting the stage for the arrival of the *Stockamsel*.[24]

Heinrich Wilhelm Döbels.

Eröffnete

Jäger = PRACTICA,

Oder

Der wohlgeübte und

Erfahrne Jäger,

Darinnen

Eine vollständige Anweisung

zur ganzen

Hohen und Niedern Jagd-Wissenschafft

in III. Theilen enthalten:

Im I. Theile wird gehandelt: 1) Von denen Eigenschafften der wilden Thiere und Vögel. 2) Wie die zur Jagd benöthigte Hunde gearbeitet, ausgeführet, eingejaget, eingehetzet, dressiret und firne gemacht werden. 3) Von Anlegung einiger Wild-Bahnen und Gehäge, insonderheit einer zahmen und wilden Fasanerey, nebst denen dazu gehörigen Sachen, und wider derselben verschiedene Krankheiten bewährten Mitteln.

Im II. Theile sind enthalten: 1) Die Jagd-Requisita, wie Stell-Flügel, Alleen, Abjagungs-Flügel, und dergleichen durch leicht von selbst zu machende Instrumente abzustecken. 2) Wie Haupt-Bestätigungs- und andere Jagen auf unterschiedliche Art einzurichten. 3) Die von Alters hergebrachte und noch in verschiedenen Europäischen Höfen floeirende Parforce-Jagd wohl zu exerciren, auch die Erziehung und Pflegung der Hunde nicht nur zu besorgen, sondern auch deren mancherley Krankheiten zu curiren. 4) Die Raub-Thiere zu tilgen, wobey zugleich die nembehelich nöthige Witterungen gezeiget werden, solche sowol, als auch alle Raub und andere Vögel, durch zur vielerley Inventiones zu fangen. 5) Das zur Jagens-Einrichtung und Fangung aller Thier und Vögel nöthige Jagd-Zeug, als Tücher, Netze, Garne, Lappen, Gärten, Gruben, Eisen, Fallen, Schlagbäume u. s. w. accurat zu machen, einzurichten und zu gebrauchen.

Im III. Theile wird vorgestellet: Die Beschaffenheit derer Holtzungen, nebst deren mancherley Benennungen, auch wie solche zu rechter Zeit und Forst-mäßig abzuholzen und abzutreiben; ingleichen deren nützlicher Verkauff und Consumirung, dieselben ordentlich zu taxiren, deren Höhe oder Länge durch leichte Instrumenten accurat zu visiren, den cörperlichen Inhalt genau auszurechnen, die abgeholtzten Oerter, dergleichen die wüsten und öden Flecken zum Zuwachs, auch durch Bepflanz- und Besaung zum Anflug und Wachsthum zu bestodern.

Nebst einem

Doppelten Anhange, 1) von besondern zum edlen Weidwerck gehörigen Wissenschafften, 2) ingleichen von der Fischerey.

Alles aus vieljähriger eigener Praxi gründlich und deutlich gezeiget,

mit vielen Kupffern und Grund-Rissen,

und einer Vorrede

Des Königlich-Preußischen Geheimen Raths und Cantzlers der Universität Halle,

Reichs-Freyherrn von Wolff.

Leipzig, 1746.

Verlegts Johann Samuel Heinsius.

Title page of Döbel's book *Jäger-Practica* (1746).

Heinrich Wilhelm Döbel

Heinrich Wilhelm Döbel (1699-1759) was a German woodsman and experienced hunter best known as the author of *Jäger-Practica*, a comprehensive treatise on hunting which was first published in 1746.

Döbel was born in 1699 in the Saxon area of Erzgebirge, the Ore Mountains that form the natural border between Saxony and Bohemia (today, it is near the border between Germany and the Czech Republic). He was the eldest of three brothers, all of whom were dedicated to hunting. His father, also named Heinrich Wilhelm, had been summoned in 1715 into the service of Prince Karl Friedrich of Anhalt-Bernburg (1668-1721) and was appointed as mounted forester in Güntersberge in Lower Harz (not far from the nearby Thuringia).

Döbel was first educated by his grandfather, Hans Rudolf Döbel, who at the time was a forester in the Ore Mountains and, after his death, by his father in Güntersberge. At that time, *par force* hunting, a form of hunting in which the game is first run down and exhausted by dogs before the kill, was flourishing in France, and the German princes imitated their example.

Döbel was determined to become an expert *par force* huntsman. He therefore traveled to various lands to perfect his *par force* hunting skills. For this purpose, he went to Potsdam, Celle, Schwerin, Nymphenburg, Ludwigsburg, Darmstadt and other places where significant *par force* events were held. After three years of wandering and well armed with experience and rich in knowledge, he returned to Güntersberge where he was appointed Huntsman in 1723 and hired in the service of Duke Ludwig Rudolf von Brunswick (1671-1735) in Blankenburg. However, he was kept busy in an office rather than being employed in the forest and this was obviously not to the liking of the young Döbel, and so he set his goal to find a job at the Dessau equipage which under Prince Leopold enjoyed a high reputation.

In September of 1725, Döbel went off to Dessau to apply for a job as a hunter. He was put to the test in a hunt where he proved his mastery and received his commission as a Piqueur (horseback hunter).

In 1733, he went to Dresden where a *par force* was set up at Hubertusburg. The head of this hunt, Prince Lichtenberg, a special favorite of the King and Elector Frederick Augustus II, recognized Döbel's ability and with the Prince's intercession he was appointed the position of Senior piqueur (Chief Huntsman) in the service of Frederick Augustus II of Saxony (1696-1763) as well as head forest ranger at the hunting lodge of Hubertusburg.

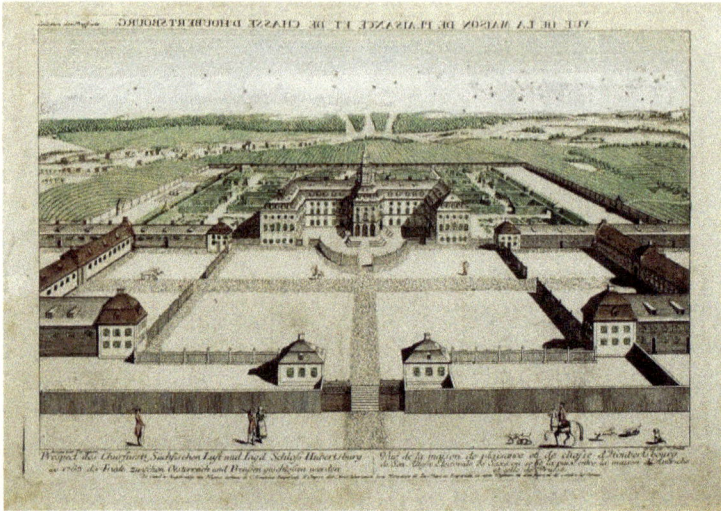

Royal Hunting Palace of Hubertusburg in 1763

Since no suitable accommodation was found for him in the hunting lodge of Hubertusburg, he took up residence in the nearby village of Reckwitz where he would write his *Jäger-practica*, the literary work that would bestow immortality to his name.

Then, the Seven-Years' War brought suffering over the Saxon Electorate and would destroy Döbel's happy life and career. It was in August of 1756 that Frederick II of Prussia invaded Saxony. The Electoral hunting was dissolved and the Royal Hunting Palace of Hubertusburg was eventually looted on January 22, 1761. This business was finished in a few hours, probably without serious damage to the building since this very castle would be the site of the Treaty of Hubertusburg at the end of the war.

In the meantime, the aged Döbel would leave Reckwitz to escape the afflictions of war and look for a new home and a new service in one of the more spared areas. Unfortunately, it is not known precisely where he went, or where and when he passed away.

CHAPTER FOUR

Kramer & the Dawn of the Stockamsel

A significant milestone in the history of the *Stockamsel* is a book titled *Elenchus Vegetabilium et Animalium Per Austriam Inferiorem Observatorum* and published in 1756 in Vienna by a certain Wilhelm Heinrich Kramer.[25] This work is a long list, with very brief annotations, of the plants and animals encountered in Lower Austria. Even though Alfred Newton in his book *A Dictionary of Birds* (1896) categorized it as "modest," Kramer's book turns out to contain the very first mention of the mysterious *Stockamsel*.[26]

> Almost nothing is known on the life story of botanist and naturalist **Wilhelm Heinrich Kramer**. His father, Johann Georg Heinrich Kramer (1684-1744), was an Austrian physician and military doctor as well as a botanical writer. In fact, the plant *Krameria triandra* was named after him. Wilhelm was born in Dresden and studied in Vienna, Austria, most likely at the Mariahilfer Gymnasium. He then officiated as town physician in the city of Bruck an der Leitha in Lower Austria, on the border between Austria and Hungary. He was married to Theresa, who bore him a daughter in 1761. Theresa died in 1763 and was buried in Bruck an der Leitha. Kramer himself died in June of 1765. In the beginning of 1766, a second daughter was born from Kramer's second wife Elizabeth.

Kramer's book is also notable for the fact that it is one of the first to employ Linnaeus's binomial nomenclature system. In fact, Linnaeus himself was so impressed by Kramer's book that, in a letter written to a colleague in 1760, he described it as "pulcherrima est" [it is very beautiful].

In his book, Kramer included a section on the birds found in Lower Austria. The *Turdus* category was listed on pages 360-361 and included the following nine birds:

1) Oriolus: *Oriole*
2) Petrocosyphos: *Great Redstart*
3) Merula vulgaris: *Common blackbird*
4) Merula torquata: *Ring ouzel*

GUILIELMI HENRICI KRAMER

Saxonis Dresdensis,
PHILOSOPHIAE ET MEDICINAE DOCTORIS

ELENCHVS
VEGETABILIVM
ET
ANIMALIVM
PER
AVSTRIAM INFERIOREM
OBSERVATORVM.

SISTENS EA IN

CLASSES ET *ORDINES*
GENERA ET *SPECIES*
REDACTA.

LABORE ET FAVORE.

VIENNÆ, PRAGÆ, ET TERGESTI,
Typis IoANNIS THOMÆ TRATTNER, Cæf. Reg.
Aulæ Typographi & Bibliopolæ, MDCCLVI.

5) Merula montana: *Mountain ouzel*
6) Turdus viscivorus: *Mistle thrush*
7) Turdus pilaris: *Fieldfare*
8) Turdus simplex: *Song thrush*
9) Turdus iliacus: *Redwing*

It is the fifth thrush on this list that interests us because Kramer also gave it the local name of *Stockamsel.*

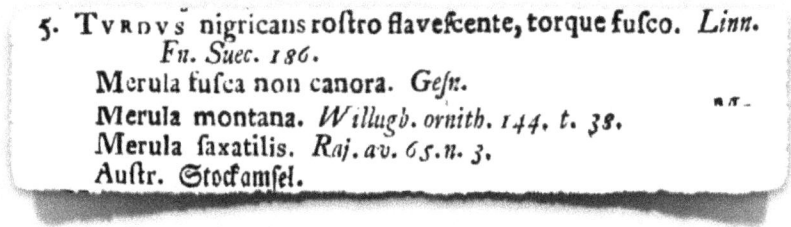

> **5. Tvrdvs** nigricans roftro flavefcente, torque fufco. *Linn.*
> *Fn. Suec. 186.*
> Merula fufca non canora. *Gefn.*
> Merula montana. *Willugb. ornith. 144. t. 38.*
> Merula faxatilis. *Raj. av. 65.n. 3.*
> Auftr. Stockamfel.

As one can see in the above image, Kramer included four alternate names for this bird:

Merula fusca non canora: from Gesner.
Merula montana: from Willughby.
Merula saxatilis: from Ray.
Stockamsel: as said by the Austrians.

Each of these synonyms is followed by the abbreviated name of an ornithology author and his book (in italics). The first three are:

a) Conrad Gesner, who introduced the descriptive term *Merula fusca non canora* in 1555.[13]

b) The reference to Willughby pertains to the *Merula montana Aldrov. saxatilis seu montana Gesneri* on page 144 of his book *Ornithologiae* from 1676, in which "table 38" contains the image of the *Merula Saxatilis Aldrov.*, also known as the *Rock Ouzell* (see page 23).[27]

c) The reference to Ray corresponds to the *Merula saxatilis* described on page 65 of his *Synopsis Methodica Avium & Piscium* from 1713.[28]

Kramer also adds a short note on this bird's habitat: "Habitat in montibus sylvaticis. Victitat iisdem"; that is, it lives in the wooded mountains and it feeds on the same food of the previously listed *Turdus torquatus* (i.e., insects and berries).

TVRDVS.

1. TVRDVS flavus; alis extremaque cauda nigris.
   Oriolus (flavus alis extremaque cauda nigris) pupillis
   rubris. *Linn. Svst. natur. 29.*
   Oriolus. *Gesn. & Willugb.*
   Chlorio. *Aristotelis.*
   Galbula. *Aldr. & Picus nidum suspendens ejusdem.*
   Austr. Engelfahraus. Norimbergensibus gelber Kirschvogel.
   Silesiis Pirohle.
   Habitat *passim in nemoribus.*
   Victitat *insectis, baccis, ficubus, & præsertim cerasis, ac pru-
   nis alias mala armoniaca dictis.*

2. TVRDVS capite cæruleo, cauda ferruginea.
   Petrocosyphos. *Gesn.*
   Austr. Steinröthl.
   an Turdus rectricibus rufis, duabus intermediis cinereis
   fascia nigricante, proxima apice cinerea. *Linn. Fn.*
   Suec. 187. pullus vel fæmina ejus? *valde enim vere
   & autumno colore differt, ac secundum ætatem & sexum
   variat.*
   Habitat *inter saxa; nidificat in lapidicinis.*
   Victitat *erucis, baccis.*

3. TVRDVS ater, rostro palpebrisque fulvis. *Linn. Fn. Suec.*
   184.
   Merula nigra. *Bellon. icon. 30. b.*
   Merula. *Aldr. ornith. l. 16. c. 6.*
   Merula vulgaris. *Willugb. ornith. 140. t. 37. Raj. av.
   65. n. 1.*
   Austr. Amsel. Amarl.
   Habitat *ubique in sylvis & nemoribus.*
   Victitat *baccis, insectis.*

4. TVRDVS nigricans, rostro flavescente, torque albo.
   Linn. Fn. Suec. 185.
   Merula torquata. *Gesn. av. 607. Aldr. ornith. l. 16. c. 11.
   Jonst. ornith. 166. t. 39. Willugb. ornith. 143. t. 37.
   Raj. av. 65. n. 2.*
   Austr. Ringlamsel.
   Habitat *in sylvis montanis.*
   Victitat *insectis, baccis.*

5. TVRDVS nigricans rostro flavescente, torque fusco. *Linn.*
   Fn. Suec. 186.
   Merula fusca non canora. *Gesn.*

Me-

Merula montana. *Willugb. ornith. 144. t. 38.*
Merula faxatilis. *Raj. av. 65. n. 3.*
Auftr. Stockamfel.
Habitat *in montibus fylvaticis.*
Victitat *iisdem.*

6. Tvrdvs pedibus fufcis, alis fubtus albentibus.
   Turdus vifcivorus. *Aldr.*
   Auftr. Zaritzer, Mißler, Zerrer.
   Habitat *in fylvis montanis, hyeme vero ubique in juniperetis.*
   Victitat *infectis, baccis.*

7. Tvrdvs rectricibus nigris, extimis margine interiore
   apice albicantibus, capite cano. *Linn. Fn. Suec. 188.*
   Turdus pilaris. *Gefn. av. 753. Aldr. ornith. l. 16. c. 1. Wil-*
   lugb. ornith. 138. t. 37. Raj. av. 64. n. 3.
   Auftr. Kranabetsvogel, Kranabeter.
   Habitat *in fylvis montanis, hyeme vero ubique in juniperetis.*
   Victitat *infectis, baccis.*

8. Tvrdvs alis fubtus ochreis.
   Turdus *fimplex. Aldr.*
   Turdus vifcivorus minor. *Bellon.*
   Auftr. Weinbrofchl, Weißbrofchl, Sommerbrofchl.
   Habitat *ubique in fylvis & nemoribus vere & æftate: autumno*
   avolat. Nidificat in fruticibus.
   Victitat *infectis, baccis.*

9. Tvrdvs alis fubtus ferrugineis, linea fupra oculos albi-
   cante. *Linn. Fn. Suec. 189.*
   Turdus illades feu Tylades, vel Iliacus. *Aldr.*
   Winfela. *Gefn.*
   Auftr. Rothbrofchl, Walbbrofcherl, Winterbrofchl.
   Habitat *circa finem autumni, & initium hyemis ubique in fyl-*
   vis noftris, tunc transmigrans, nec vere quamquam iterum
   tranfeuns apud nos nidificat, jam autem in Bohemia.
   Victitat *infectis, baccis.*

List of the Turdi in Kramer's *Elenchus Vegetabilium et Animalium Per Austriam Inferiorem Observatorum* (1756).

The most interesting name on Kramer's list is the last synonym: *Stockamsel*. This appears to be the first mention of such a bird name, a term that Kramer tells us originated in Austria. The word *Stockamsel* derives from the German words *Stock* ("stump") and *Amsel* ("ouzel"). This is also the first time that the *Stockamsel* will be categorized as a separate and distinct type of bird, here as one out of nine types of thrushes. Kramer notably separated it from both the *Blackbird* and the *Ring ouzel*. We shall now examine in a little more detail the four bird synonyms that Kramer identified with the *Stockamsel*.

(1) *Turdus nigricans, rostro flavescente, torque fusco* of Linnaeus

In 1746, **Linnaeus** published his *Fauna Svecica* in which he gave a list of birds in the Turdus genus that included: "186. TURDUS nigricans, rostro flavescente, torque fusco" [#186. Blackish thrush, yellowish beak, brown neck ring].[29]

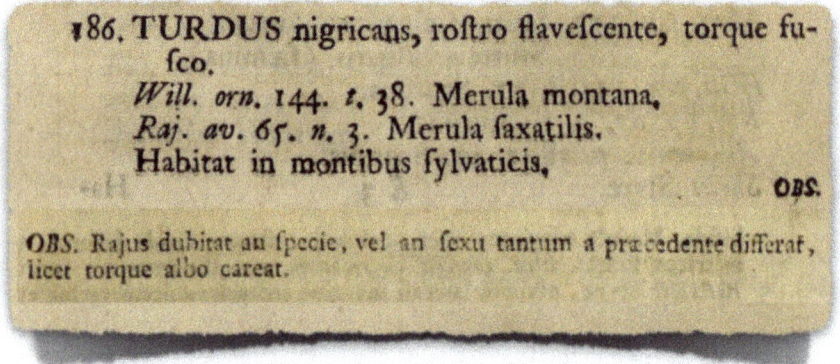

> **186. TURDUS** nigricans, roſtro flaveſcente, torque fu-
> ſco.
> *Will. orn.* 144. *t.* 38. Merula montana.
> *Raj. av.* 65. *n.* 3. Merula ſaxatilis.
> Habitat in montibus ſylvaticis.
>   **OBS.**
>
> **OBS.** Rajus dubitat an ſpecie, vel an ſexu tantum a præcedente differat,
> licet torque albo careat.

Linnaeus associated this bird with two bibliographical references and their respective synonyms, one by Willughby and one by Ray, with the same book references and page numbers cited by Kramer.

(2) *Merula montana* of Willughby

The *Merula montana* was described by the English ornithologist Francis **Willughby** in the 2nd volume of his book from 1676 called *Ornithologiae*.[27] On page 144, Willughby called it the *Merula montana Aldrov. saxatilis seu montana Gesneri* ["Mountain ouzel of Aldrovandi Rock or Mountain ouzel of Gesner" – a confusing name, two years later translated into English by John Ray as "Rock ouzel or Mountain ouzel of Gesner"]. An image of the *Merula Saxatilis Aldrov.* or the *Rock Ouzell* is depicted on his table 38 (see below).

Willughby's woodcut image of the "*Rock ouzell*" corresponds to the image that Kramer referred to when mentioning the *Stockamsel,* and thus would appear to be the very first depiction of the *Stockamsel* itself.

And yet, this drawing is very similar to, and most likely a mirror image of, the woodcut image of the *Merula saxatilis* or *Caryocatacte* depicted in the book *Ornithologiae* by the Italian naturalist Aldrovandi in 1600.[30] This bird's name has caused much confusion. Aldrovandi gave it additional synonyms: *Corresolo* by the Italians, *Merle alpadie* or *Merula alpina* in the Alps and near Lake Verbano, *Cassenoix* by the French, *Nucifragum* in Latin, and *Nutzbrecher* by the Germans.

Aldrovandi's illustration and, consequently, that of Willughby do somewhat resemble a *Spotted Nutcracker* (*Nucifraga caryocatactes*). This bird has an extensive range of distribution, from Scandinavia, northern Europe, through Siberia, to East Asia and Japan. It inhabits the mountainous regions of the Alps, the Carpathians, and the Balkan Mountains. In contrast, it is quite rare in England, as opposed to the many sightings of the so-called *Rock ouzel*. These two images would lead one to surmise that the *Stockamsel*, the *Rock ouzel* and the *Nutcracker* may be one and the same. Future evidence by ornithologists would

The Eurasian *Spotted Nutcracker*.

refute this and Willughby's woodcut image turns out to be neither a *Stockamsel* nor a *Rock ouzel* and not even a *Merula saxatilis*, but seemingly that of a *Nutcracker*. Gesner first described the *Caryocatacte* in 1555 and called it also *Merula alpina* – but not *M. saxatilis* as Aldrovandi did. We can rightfully blame Aldrovandi for initiating this confusion.

(3) *Merula saxatilis* of Ray

Kramer used the term *Merula saxatilis* as written by the English naturalist John **Ray** in his book *Synopsis Methodica Avium & Piscium* (1713).[28] Ray also called it a *Rock Ouzel*, but the nature of this bird is shrouded in doubt since even Ray demonstrated a lack of confidence regarding its uniqueness. In fact, he said that, even though the *Rock ouzel* lacked the neck collar, it resembled the *Ring ouzel* and doubted whether it was truly distinct from the *Ring ouzel*. Ultimately, the *Merula saxatilis* of Ray would end up as either the *Merula montana*, the *Bergamsel* or the *Rock ouzel* of other authors. And the *Rock ouzel*, especially, would often be equated with the juvenile and female *Ring ouzel*.

A.3. M ERUL A *faxatilis. The* R O C K O U'Z E L. Præcedenti fimilis eſt, ſed torque caret. In montofis *Pecci Derbienſis* propè vicum *Hatherſ⸿dge* dictum in præruptis *e/* rupibus è quibus lapides molares exfcinduntur vidimus plures. Defcriptionem *vide* in Ornithologia *D. Willughby. It is frequent on the high Mountains of* Caernarvanfhire *and* Merionydhfhire *in* Wales, *where they call it* Mwyalchen y graig, *quaſi dicas*, Merula rupicola. I *have alfo feen feveral of them on a high Hill in* Ireland, *called by the Englifh*, Mount Lemfter, *in Irifh* Stoa Lein. *D. Lloyd.* An hæc à præcedente, fpecie, an fexu tantùm diftincta fit, aliquando dubitavi, necdum mihi plenè fatisfit.

(4) *Merula fusca non canora* of Gesner

The last synonym used by Kramer is the *Merula fusca non canora* of Gesner. In 1555, the Swiss naturalist Conrad **Gesner** wrote in his *Historiae Animalium* that the *Merula torquata* ["ring ouzel"] was called in Switzerland as *Birgamsel* ["mountain ouzel"] as well as *Steinamsel* ["rock ouzel"], among others.[13] He went on to say that the *Birgamsel* is also the name for the *Merula fusca non canora* ["non-singing fuscous blackbird"]. When describing the three *Ouzels* of Aristotle, of which the first was the common *Blackbird* and the second a white variant of *Blackbird*, Gesner ascribed the term *Merula fusca* to Aristotle's third kind of *Blackbird* but without any mention of its singing capabilities. In fact, while giving a detailed description of the *Merula fusca* from his own personal observations and stating that its beak was black and the belly black admixed with grey, Gesner failed to mention the adjective 'non canora.'

It would be in his bird atlas *Icones Avium Omnium* from 1555 that Gesner would add the attribute *non canora* to the *Merula fusca*.[31] He wrote:

> But the latter name [that is, *Birgamsel*] is often attributed to the type of blackbird that is dusky and non-singing ("fusco, non canoro").

Unfortunately, the *Merula fusca non canora* and its other alternate names did not have a successful run and would ultimately be considered as the no less mysterious *Bergamsel* or as a variant of the *Ring ouzel*, or even as the female *Blackbird*.

> Note: "fuscus/a" can be translated as fuscous, dusky, dark, or brown.

It should be noted that Kramer used the term *Stockamsel* instead of *Stockziemer*, a term introduced by Döbel a decade before. At the time, the *Amsel* (*Merula* in Latin) and the *Ziemer* (*Turdus* in Latin) were considered two distinctly different birds. The *Amsel*, also known as *Blackbird*, or *Merle* in French, corresponded to the thrush-type of blackbird in which the colors of the breast plumage were either uniform or distributed in large masses. The *Ziemer*, on the other hand, implied a thrush-type of bird with a spotted, dappled or speckled breast. The latter is given the English appellation of *Thrush*, in Latin *Turdus*, and in French *Grive* (from the French *grivelé* or speckled). The reason for the dichotomy of terminology between *Stockamsel* and *Stockziemer* is never explained by any author, neither then nor in the future, and these bird names will basically run on parallel tracks with no one attempting any comparison between the two, other than mentioning both as synonyms of the *Ring ouzel*.

In summary, the *Stockamsel* made its first grand appearance with Kramer but was unfortunately associated with bird types, none of which would stand the test of time. The *Turdus nigricans, rostro flavescente, torque fusco* of Linnaeus lasted only two editions of his *Fauna Svecica* (1746 and 1761) and, under the name of the *Merula montana*, it never made it beyond the 9th edition (1756) of Linnaeus's *Systema Naturae*.[29,32,33] And both Willughby and Ray were dubious of the uniqueness of the *Mountain ouzel* and felt that it was nothing more than a variant of the *Ring ouzel*.

We must not lose sight of Kramer's strategic placement of the name *Stockamsel*, i.e., completely separate from both the *Turdus merula* and the *Turdus torquatus*. He equated it with the *Merula fusca non canora* and the *Rock ouzel*, two birds with a long and complex history. It can be anticipated that this will be the only time that the *Stockamsel* will be associated with these two birds, for we will see that almost every author from now on will consider it as equivalent to either the *Blackbird* (*Turdus merula*) or the *Ring ouzel* (*Turdus torquatus*).

After Kramer's introduction of the *Stockamsel*, there would be a 40 year-long period of silence on the pages of Natural History books regarding this bird. It would not be until 1795 that another author, the great ornithologist Bechstein, would re-introduce the term *Stockamsel*.

The *Stockziemer*, meanwhile, would be awarded the honor of representing an elusive ouzel hypothetically related to tree stumps. During the latter half of the 18th century, it received less than satisfactory attention and was predominantly mentioned in hunting and forestry books without much description nor any first-hand evidence. The name *Stockziemer* would be employed by various authors, all of whom would associate it either directly or indirectly with the *Ring ouzel*. It was often equated with the *Meeramsel*, a synonym for the *Ring ouzel*, even though some authors such as Bechstein would associate it also with the *Stockamsel*. However, its course had apparently been set and the ornithology world would essentially equate it with the *Ring ouzel*.

CHAPTER FIVE

The Era of Count de Buffon

We now turn our attention to Kramer's neighboring country France. Even though the *Stockamsel* never made a dent in the wisdom of French ornithologists, the confusing plumage of blackbirds did create a new spin-off called *Merle brun* ["Brown ouzel"].

It all began in 1767 when the physician Dr. M. **Salerne** authored the book *L'Histoire Naturelle eclaircie dan une des ses partie principales, L'Ornithologie*, which is basically the French translation of Ray's *Synopsis Avium* augmented by many descriptions and remarks by Salerne.[34] When discussing the *Blackbird*, he said that the most well-known species of *Blackbird* in Germany is the one whose male is black with a waxy yellow beak; but the female is completely different and is considered by some as a bird of another species ["autre espèce"]. The expression "autre espèce" most likely refers to another thrush-type of bird.

A second passage in Salerne's chapter on the *Blackbird* may be of relevance. Here he discusses the implication of differing nesting habits among the *Blackbirds*. Salerne states that there are those who claim that two kinds of Blackbirds exist; namely, the ordinary or common *Blackbird* that nests at a certain height in trees or shrubs, and the *Merle terrier* ["Ground blackbird"] which makes its nest low and on the ground. Salerne claims that this distinction is not justified and tells the story of a *Blackbird* that, after a cat had eaten her first two broods in a nest at the foot of a hedge, made its third nest on an apple tree at the height of about eight feet.

Merle terrier

In a 1901 article for the journal *Le Cosmos*, Armand Jean **Leray** agreed with Salerne and stated that if the so-called *Merle terrier* had at its disposition, in the month of March, a box hedge or a well-stocked wall of thick ivy, it would prefer to hide its nest in the evergreen foliage.[35] Confusingly, Salerne himself, in his same book, also regarded the French term *Merle terrier* as a synonym for the *Ring ouzel*. This would be repeated by Buffon in 1775, and many others would follow suit. On the other hand, the analogous term *Merlo terragnolo* would be given to the 'stockamsel' variant of the *Blackbird* by Arrigoni degli Oddi in 1902.[36] Salerne did warn us against creating new bird species based on their nesting habits.

The notion of the female *Blackbird* as representing another species would appear in the very popular *Histoire Naturelle des Oiseaux* [Natural History of Birds] written in 1775 by George-Louis Leclerc, Comte de **Buffon**.[37] Neither the *Stockamsel* nor the *Stockziemer* made an appearance in his book, but Buffon did introduce a variant of the *Blackbird* that may well be equivalent to the *Stockamsel*. His chapter on the *Blackbird*, which was actually written by Guenau de **Montbeillard**, stated that the female *Blackbird* had no black whatsoever throughout its plumage; instead it showed different shades of brown mixed with red and grey, while its beak was only rarely yellow and it did not sing as well as the male. All this led many to consider it as a bird of another species ["autre espèce"]. Again, we have the allusion to an unnamed "other species," but this time, hidden in a footnote, we are given a clue to the identity of this elusive "species." In fact, Montbeillard refers the reader to figure 29 (see next page) in a book recently published by the German ornithologist Johann **Frisch**, *Vorstellung der Vögel Deutschlandes* [Introduction to the Birds of Germany] (1740), and then adds that he suspects that in certain lands the name *Merle-grive* was given to this female bird.[38] Buffon's book is the first in which a specific bird (i.e., the *Merle-grive*) is accused of being the infamous "autre espèce" that is ultimately none other than the female *Blackbird*. We will later see how the *Stockamsel* would inherit the same role.

Montbeillard's reference to **Frisch** had to do with a passage in this German author's chapter on the *Blackbird* wherein he said that the female looks much different than the male – as can be seen in the colored figure (see next page) – and is believed by some to be a different type of bird and classified among the *Thrushes*. This passage is notable because Frisch emphasized the dilemma surrounding the female *Blackbird* which some have considered a thrush-type of bird due to its spotted breast.

In the 1800 edition of Buffon's *Histoire Naturelle des Oiseaux,* the French naturalist **Sonnini** would elaborate on the *Merle-grive* (which he also called *Merle brun*).[39] To the sentence of Montbeillard that the female *Blackbird* is confused by some as a bird belonging to a different species, Sonnini retorts that he does not share in Montbeillard's opinion and, along with all those who have observed the blackbirds of France, he is certain that there are two species that are distinct and consistently different from each other. One is the common *Blackbird*, and the other

The female and male Blackbird depicted in *Vorstellung der Vögel Deutschlandes* (1740) by Johann Leonard Frisch.[38]

is the *Merle brun* [brown blackbird] or *Merle-grive* [blackbird-thrush]. Sonnini states that it is in the districts covered by forests and filled with blackbirds, such as Lorraine, where one can easily be convinced of the truthfulness of this distinction — a distinction well known to the local bird-catchers. To support the separation of these two species Sonnini reports on the observations made by a fellow ornithologist, Monsieur **Tillancour**, one of his old neighbors on the border between the regions of Lorraine and Champagne. In a letter to Sonnini, Tillancour confirmed that the adult *Blackbird* is black except for its beak, its eye-rim, and the heel and ball of the foot, which are more or less yellow. The female, however, has no pure black anywhere in its plumage, but only different shades of brown mixed with grey, and only rarely has a yellow beak. This disparity has caused some to consider the female a bird of another species, which is referred to in some areas as *Merle-grive*. Tillancour emphasizes that Guenau de Montbeillard wants to make a single species of all the blackbirds nesting in France, both those *with a yellow beak* (i.e, common Blackbird) and those that are *grey, red or brown*, which he believes are either female or juvenile blackbirds or display such color as a result of moulting. According to Tillancour, Montbeillard is mistaken in this regard because the *grey-red-brown* species is not the same as the black species *with a yellow beak*. His contention of a marked difference between these two species is based on three facts: (1) *Blackbirds with a yellow beak* spend the winter in Lorraine and are more resistant to the cold, while the *Merles bruns* do not stay there much and many have died during the heavy snows; (2) one captures in the nets a much larger number of *Merles bruns* than *Blackbirds with a yellow beak*, especially in October during the time of migration, well after the moulting season when the adult-like young *Blackbirds with a yellow beak* of that year have nearly as much yellow in their beak as the adult; and (3) the *Merles bruns* are larger than the *Blackbirds with a yellow beak*, which have a smaller and more compact body.

These circumstances had led Tillancour to erroneously conclude that these blackbirds were of a different species. Tillancour added another feature that emphasized the difference between these species: the male *Blackbird with a yellow beak* easily learns to whistle in a cage and even to speak, something that the males of other species cannot do so easily and so well. Thus, when raising young blackbirds, this difference in singing makes one always prefer those with a yellow bill. When bird hunters find a *Blackbird* nest and realize that the parents are just brown, they do not believe that the young birds are worth raising. But when they realize that the parents are black and, above all, that their beak is yellow, that is when they think they have made a real find. So, for Tillancour, this was further evidence that *Blackbirds with a yellow beak* form a separate species or race and are distinct from other blackbirds.

Due to the words of his friend, Sonnini adamantly considered the *Merle brun* (or *Merle-grive*) as a distinct species of *Blackbird*, in direct opposition to the view expressed by Guenau de Montbeillard, author of the very chapter that Sonnini is commenting on. The letter that Sonnini received from Tillancour is also of interest since it opens a window into the world of 18th century ornithology. Not only did ancient ornithologists such as Sonnini obtain their information from the previous literature on the subject as well as from their own personal observations, most of it through bird hunting, but also from friends and fellow bird enthusiasts in differents parts of the country or abroad with which they exchanged information. A constant source of material was from hunters or bird catchers. Another source was the local market where birds, whether caught alive for breeding or to be kept as pets, or killed for culinary consumption, were exhibited.

Even though Tillancour made a passionate case for the separation of the *Merle brun* (or *Merle-grive*) from the *Blackbird*, which is otherwise called *Merle au bec jaune* [blackbird with a yellow beak], most of his arguments can be used against him to prove that the *Merle brun* is nothing more than a female and/or a first-winter young male *Blackbird*. For instance, his first argument may be dependent on the survival of the strongest and fittest, and adult male *Blackbirds* are at an advantage over younger individuals. Secondly, the sum of female and first-year *Blackbirds* surely must outnumber adult males, and this would explain their higher number being caught in the nets. And thirdly, the authors of the time consistently stated that the female *Blackbird* was larger than the male. To some extent, the same holds true today. Simms said that females weigh a little less than males but that there is also significant overlap in their weights.[40] Stephan showed that the average weight of the female is greater than that of the male, especially in spring [87 vs. 83 gm]; while the opposite is true in autumn [92 vs. 95 gm].[6]

The *Merle brun* did make an appearance in the German language literature. In fact, a second German edition of **Buffon**'s book was published in 1784 under the title *Naturgeschichte der Vögel*, with comments added by Bernhard Christian Otto.[41] He wrote that some hunters claimed that there were two types of *Blackbirds*; a *black* one, and a slightly *larger brown* one, the males of which always remained brown. He found that the latter males had a black-brown back, a whitish lower neck flecked with brown, a rusty reddish-brown breast, and a grey-brown abdomen. They were also somewhat larger than the *Blackbird*. These were similar to the female *Blackbird* and were only recognized as males by the presence of testicles. Nevertheless, Otto believed that they were probably males which were not yet completely adult. This brown variant of *Blackbird* (or *Great brown ouzel*) described by Otto strongly evokes the French *Merle brun*. He mentions how these are difficult to distinguish from the *Blackbird*, especially the female bird, and with foresight suggests that they are merely first-winter male *Blackbirds*.

After Frisch and Buffon, the issue of the *Merle-grive* will come up again, with opposite interpretations. In fact, in 1795 Johann A. E. **Goeze** discounted the *Merle-Grive* as just another name for the female *Blackbird*.[42] On the other hand, Jacques-Joseph **Baudruillart** wrote in 1834 of the *Blackbird* and said that the female differs so much from the male to the point that one would consider them as two different species of bird.[43] He then added a section where he said that the *Merle Brun* is a species, known also as *Merle-grive*, which differs from the common *Blackbird* by its slightly larger size and its constantly brown beak. He stressed how it had been confused, and still is confused, with the female yellow-billed *Blackbird*, and that it is common in Lorraine. Fully recognizing that the female *Blackbird* due to its thrush-like plumage can be confused with another species, Baudruillart still insisted that the *Merle brun* or *Merle-grive* was a separate species.

In 1823, the French abbot and naturalist **Bonnaterre** and the ornithologist **Vieillot** co-authored the *Tableau Encyclopédique et Méthodique des Trois Règnes de la Nature - Ornithologie*.[44] Upon discussing the *Blackbird*, they said that there exists a type whose plumage is analogous to that of the female *Blackbird* and is known as *Merle brun* [brown ouzel] and *Merle gris* [grey ouzel]. This ouzel is distinguished from the *Blackbird* by its larger size, its constantly brown beak, and by the lack of ease it shows in learning to talk and whistle when in captivity. Bonnaterre's description of this variant of *Blackbird* is remarkably reminiscent of the *Stockamsel*.

Merle Gris

The *Merle gris* suffered a fate similar to that of the *Stockamsel* by being tossed back and forth between the *Blackbird,* the *Ring ouzel* and others.

• In 1814, Comte **de Bray** wrote in his *Mémoire sur la Livonie* that the *Merle gris* was the *Turdus musicus* (Song thrush).[45]

• In 1826, Auguste **Drapiez** wrote in the *Dictionnaire Classique d'Histoire Naturelle* that the *Merle gris* was synonymous with the *Ring ouzel.*[46]

• In 1836, the book *Secrets, Anciens et Modernes, de la Chasse aux Oiseaux* gave it a description that perfectly matched that of the *Ring ouzel.*[47]

• In 1868, J. M. **Guardia** upon reviewing the book *Le Victorial, Chronique de don Pedro Niño* stated that the *Merle gris* was the *Turdus merula.*[48]

A rather crucial statement regarding the essence of the *Merle brun* appeared in the entry on the "Merle" in volume XXX of the *Dictionnaire des Sciences Naturelle* written by Charles **Dumont** in 1824.[49] He ended the section on the *Merle commun* (Blackbird) with the following paragraph:

> Certain Naturalists claim that there is a particular breed called *Merles bruns* [brown blackbirds]; but it is likely that the birds considered as such are nothing more than females, or young birds who had been slower (or delayed) in acquiring the adult plumage.

This is the first time that the plumage of the *Merle brun* (a French bird that would appear to correspond to the German *Stockamsel*) is explained as that of young birds "lingering" or being late when moulting into the adult plumage. Dumont introduces here the notion of "delayed" acquisition of the adult plumage. This would be resurrected a century later as the concept of "Hemmungskleid" (retarded plumage) by Erwin Stresemann in his monograph *Avifauna Macedonica* (1920) and would be advanced as the most likely explanation for the plumage of the *Stockamsel*, well into modern days.[50]

A search for the *Stockamsel* in the British ornithological literature of the time fails to reveal any sightings. One might hope to find a mention of the *Stockamsel* in the popular treatise *A History of British Birds* written in 1797 by Thomas **Bewick** and Ralph Beilby but all one finds is a passage written in the chapter on the *Blackbird*:

> The female is mostly brown, inclining to rust colour on the breast and belly; the bill is dusky, and the legs brown; its song is also very different, so that it has sometimes been mistaken for a bird of a different species.[51]

From the way Bewick expresses himself, it would seem that the female *Blackbird* is mistaken for another species not so much for its appearance but rather for its song. It is thus impossible to make any analogy with other works.

Female *Blackbird* as depicted in Thomas Pennant's *British Zoology* (1776).[52] Notice how the chest and abdomen display some pale grey feather edging. This feature is also present in first-winter male *Blackbirds* as described by Seebohm and Witherby, but even more so in the *Stockamsel* as mentioned by Vinicombe et al.

CHAPTER SIX

Bechstein

After Buffon's epic work, the ornithology literature continued unperturbed on its path of repeated citations of the *Stockziemer* and persistent confusion generated by the plumage of the female *Blackbird*. Buffon's "autre espèce" continued to remain vague. The *Blackbird* was not the only species to be considered a "confusion species" – one in which its two genders are considered by some to represent two separate species. The *Ring ouzel* (*Turdus torquatus*) would suffer the same fate. Indeed, in 1775, Rev. Samuel **Ward**, when discussing the *Ring ouzel*, stated:

> The breast is adorned with a white crescent in the middle, with the horns pointing to the hind part of the neck. This crescent is of a pure white in some, and of a dusky hue in others. Neither the females nor any of the young birds are possessed of this mark, which has occasioned some naturalists to form two species of them. [53]

Of course, this is not exactly true, because the females and the young are the ones with the crescent "of a dusky hue." It might be that the author was referring to the *Rock ouzel* as the mistaken second species. Rev. Ward does raise the issue that the *Ring ouzel*, especially the young and the female adult, is a confusion species and ultimately a possible contender in the race to be the real *Stockamsel*.

The end of the 18th century would not only see more citations of the *Stockziemer*, such as in the dictionaries of Adelung[54] and Nemnich,[55,56] both of which defined the *Stockziemer* as the *Ring Ouzel*, but most importantly it would give the ornithological world the first published work of the great Bechstein who struggled more than a little with the concept of the *Stockamsel*.

It was in 1795 that Johann Matthäus **Bechstein** published the fourth volume of the first edition of his *Gemeinnützige Naturgeschichte Deutschlands nach allen drey Reichen* [General Natural History of Germany covering all three Kingdoms].[57] As synonyms of the *Ring ouzel* he included the names *Bergamsel* and *Stockziemer*, as well as *Stockamsel*, the name by which it was called in the region of Thuringia in Germany. The only other previous mention of the latter term was by Kramer in 1756 as a synonym of the *Merula montana*. Besides being equated with the *Stockziemer*, this is the first time that the term *Stockamsel* is applied directly to the *Ring ouzel*, though without any explanation. In the description of the copper-plate figures at the beginning of his book, Bechstein equated the *Stockamsel* with both the *Ring ouzel* and the *Blackbird* when he said that local hunters and bird-catchers always speak of the *Stockamsel*, a name they give to a type of thrush reminiscent of the *Blackbird*. On closer examination, Bechstein found that they meant either a young *Blackbird* or the *Ring ouzel*. He then fatefully referred the reader to a picture of a bird which was none other than a classic male *Ring ouzel* (see below). And so, for years to come, ornithologists would confusingly consider the *Stockamsel* as a *Ring ouzel* even though others preferred to equate it with the *Blackbird*.

Tab. IV.

In the chapter on the *Blackbird*, Bechstein did not include the term *Stockamsel* as a synonym. However, as others have previously done, he described the female *Blackbird* as brown-black with a rust-colored breast, ash-grey belly, yellow inside of the bill, brown-black plumage, and a pale throat speckled with dark brown. It also appeared to be somewhat larger and heavier than the males. For these reasons, the local hunters have wanted to make of her a special kind of blackbird, and they would bring Bechstein this "brown-black" ouzel and call it with another name. Bechstein does not immediately tell us what this name was, but when mentioning several variants of the *Blackbird* at the end of this chapter, he added in a footnote:

> The *schwarzbraunen* [Brown-black ouzel], which sometimes is thought to be its own species, is nothing more than the female (Blackbird).

Bechstein expands on this statement in this volume's appendix:

> Here I have remarked in the note that the *Brown-black blackbirds*, which are called by fowlers and bird enthusiasts as *Stockamsel* and *Meeramsel*, are nothing more than female [Blackbirds].

Meanwhile, Bechstein had realized that well-fed young birds are usually somewhat larger than those living in the free, and that they also have the colors of the female. He was doubtful whether this was a constant variant of *Blackbird*, let alone its own species, even though the fowler was convinced of this. He then relayed what a local amateur ornithologist and friend, Mr. von **Schauroth**, had to say:

> Here the fowler makes a great distinction between the *Blackbird* and the *Stockamsel*. Of the latter, I have had an old lame male and a young one. These have only one bright yellow beak, eyelids not bright yellow but almost white, and the plumage dull black, which is due to the fact that all the feathers have a subtle grey or brown edge which is most evident on the wings, and the shoulders are dusted with grey edges. The young birds resemble the *Blackbird*, though they also have the feather edging, and already learn to sing. They have a much stronger whistle, almost like the Orioles, and place their nests on the ground or on a tree stump. My old injured bird (which a fowler sent me, whom you certainly would want to meet) did not live long, but the young one that I had, sang quite pleasantly like the others.

Bechstein has thus offered an in-depth discussion of the *Stockamsel* or *Schwarzbraunen amsel* [Brown-black ouzel], otherwise known as *Meeramsel*. He was not convinced of the *Stockamsel* as an entity unto itself but nonetheless offers the anecdotal arguments of the locals. Of the utmost importance is the announcement by his friend, Mr. von Schauroth, of the nesting site of the *Stockamsel* – "on a tree stump" – a significant detail that would justify the bird's singular name.

Elsewhere, such as in the chapter on the *Blackbird*, Bechstein denies the *Stockamsel* any particular separate status and considers it a form of *Blackbird*. In fact, he specifically says that the female and young male *Blackbirds* are referred to as *Stockamsel, Graue amsel* [Grey ouzel],

Graudrossel [Grey thrush], *Braun merle* [Brown blackbird], or *Bergamsel* [Mountain ouzel].

Thus, within the very same book Bechstein speaks of the *Stockamsel* as a variant of both the *Blackbird* and the *Ring ouzel*.

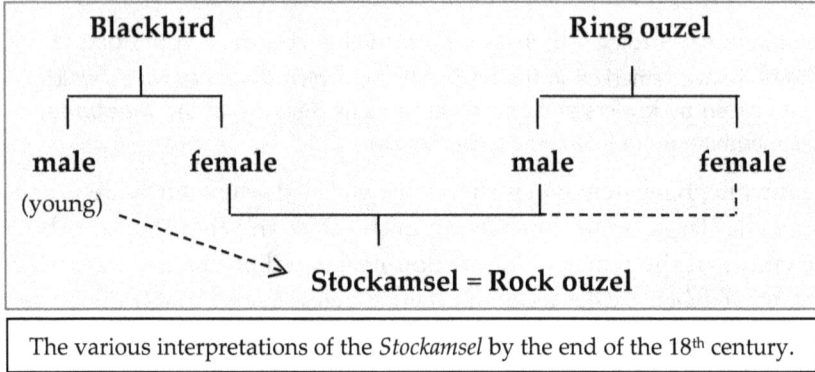

The various interpretations of the *Stockamsel* by the end of the 18th century.

Johann Matthäus Bechstein (July 11, 1757 – February 23, 1822) was a German naturalist, forester, ornithologist, entomologist, and herpetologist. He was the son of Andreas Bechstein, a blacksmith and armourer, and was born in Waltershausen in the district of Gotha in Thuringia. He studied theology for four years at the University of Jena and spent time hunting and roaming the forests as opportunities permitted. After leaving school, Bechstein taught for some years, but gave up teaching to devote himself to outdoor pursuits. In 1795, he founded the school of forestry at Waltershausen, and in 1800 the Duke of Saxe-Meiningen made him the director of the forestry school at Dreissigacker near Meiningen in the neighbouring district of Schmalkalden-Meiningen. He remained in the service of this prince till his death in 1822. After the death of his own son, Bechstein adopted his nephew Ludwig Bechstein.

That same year, Johann **Goeze** published the book *Europäische Fauna oder Naturgeschichte der Europaischen Thiere*.[43] He, too, mentions the synonyms of *Bergamsel* and *Stockamsel* when talking about the *Ring ouzel*. Goeze would be the third author to mention the term *Stockamsel*. He chose to associate it with the *Ring ouzel*, as Bechstein did in his prologue, in contrast to what Bechstein wrote in his body text.

Bechstein would repeat the synonyms *Stockamsel* and *Stockziemer* for the *Ring ouzel* in his 1797 book *Gründliche Anweisung alle Arten von Vögeln*.[58] In the chapter on the *Blackbird*, he replicates what he previously wrote about the female *Blackbird*, but this time Bechstein continues:

> It is also said that the *Stockamsel* is a distinct species. But I have noticed that these are mostly young reared *Blackbirds* which immediately after the first moult, or when they are several years old, acquire such a color. They are, namely, smoky black, slightly lighter underneath; they have a whitish throat with long dark brown stripes, and a beak that is half black and half yellow. This type is said to be much more graceful, but this is quite obvious since they have practiced their singing with several melodies while in captivity.

In this book, Bechstein tackles the essence of the *Stockamsel* with more directness and gives us one of the better descriptions of this alleged species.

Bechstein's future books will continue to give the synonym of *Stockamsel* (and that of *Stockziemer* and/or *Bergamsel*) to the *Ring ouzel*. He reemphasizes that the name *Stockamsel* (and *Bergamsel*) is also applied by bird enthusiasts and fowlers to young male blackbirds raised in captivity, and that in the wild they can be found as females of the *Blackbird*.

In the second edition of his *Gemeinnützige Naturgeschichte Deutschlands* (1807), Bechstein continues to state that the female *Blackbird* is called by local hunters with the name of *Stockamsel* or *Bergamsel*.[59] It is at this point that he adds a reference to a book by Wirsing et al (1772) and its figure no. 23 (see next page) which depicts a bird with the caption *Bergamsel* and *Merle de montagne* (i.e., "Mountain ouzel"). Wirsing's book includes additional synonyms for this bird such as *Merula montana*, *Merula tertii generis*, and *Rock ouzel*. Wirsing's text-atlas *Sammlung meistens Deutscher Vögel* was published in 1772 in Nuremberg as a collaborative effort by three individuals who contributed the following: paintings by Barbara Regina **Dietzschin**, engravings by Adam Ludwig **Wirsing**, and descriptions by Dr. Benedict Christian **Vogel**.[60] The book's text is

The "Bergamsel" in Wirsing's *Sammlung meistens Deutscher Vögel* (1772).

taken from several sources, especially Brisson and Gesner, and the authors make no mention of the *Stockamsel* itself within their description of the *Bergamsel*. It is unclear if the figure depicts an actual bird specimen or an imaginary one created from the written description itself. Even so, the bird depicted does not have much of a pale throat, the distinguishing feature particularly stressed by Bechstein.

Bechstein also restated the fact that the *Stock-amsel* or *Bergamsel* sing better, but only because their song is enriched while in captivity. By attributing their song quality to the artificial environment in which they are raised, this last passage eliminates one of the features that differentiate the *Stockamsel* from the *Blackbird*. In his book on the natural history of caged animals, Bechstein goes into detail on the motives that brought the *Stockamsel* to the attention of birders.[61] The young male *Blackbird* was eagerly sought for due to its ability to sing, and especially for the ease with which one could teach it to sing while in captivity. It was while in this state of captivity that the birder would realize that a certain number of these *Blackbirds* did not develop the coal black plumage of the typical adult. These were thus called *Stockamsel*, a name that had already been granted to a particular type of *Blackbird* known to nest in tree stumps.

In 1838, **Bechstein**'s *Naturgeschichte det Stuben-Vögel* was translated into English as *The Natural History of Cage Birds*.[61] The latter book, as well as Bechstein's *Gemeinnützige Naturgeschichte Deutschlands nach allen drey*

Reichen, were well read by the famous Charles **Darwin** who made many references to them in his book *The Variation of Animal and Plants under Domestication* (1868).[62] Darwin commented on the change in plumage of birds in captivity, and would reference Bechstein when discussing the influence of food, light, and captivity on the plumage of cage-birds:

> For many cases have been recorded of the loss by male birds when confined of their characteristic plumage.

> But in all these cases the food probably is much less varied in kind than that which was consumed by the species in its natural state.

> Bechstein does not entertain any doubt that seclusion from light affects, at least temporarily, the colours of cage-birds.

Here we have another possible explanation for the plumage of the *Stockamsel,* as seen in other caged birds. Both Bechstein and Darwin suggested a link between the change in diet and lack of natural light on the one hand and the change in plumage color of birds in captivity on the other. Poor diet in birds has indeed been proposed as an explanation for the presence of white, melanin-deficient feathers in certain species. Darwin was so impressed by Bechstein and his books that he even claimed that Bechstein was "the highest authority on cage-birds."

And so ends the first quarter of the 19th century which was dominated by the presence of Johann M. Bechstein and the plethora of his natural history books. His simultaneous attribution of the *Stockamsel* to both the *Ring ouzel* and the *Blackbird* would trigger a rash of confusion that would permeate the ornithological literature for decades to come. There was certainly no air of confidence in Bechstein's 1820 book where he simultaneously categorized the *Stockamsel* as a *Ring ouzel,* a female *Blackbird,* and a young male *Blackbird.*

CHAPTER SEVEN

Naumann

The late 18th century ornithological scene would experience the arrival of the soon-to-be famous Johann Andreas **Naumann** who, in 1789, wrote his first ornithological book *Der Vogelsteller* [The Bird Trapper].[63] Within the category of the "Amsel" he included the *Ringamsel* (Ring ouzel), the *graue Amsel* (Grey ouzel) and the *schwarze Amsel* (Black ouzel). Of the *graue Amsel* (which he will later equate with the *Stockamsel*) Naumann writes that it is slightly smaller than the *Ring ouzel*, appears grey-black, and its breast is ash-grey mixed with brown. Even though similar in size, the *Blackbird* differs because of its pitch-black feathers and yellow beak. They both have similar habits and voice, except that the *Grey ouzel* is not as good a singer as the *Blackbird*. With the alternate name of *Graue-amsel*, this is technically the second appearance of the *Stockamsel* in the ornithological literature, preceding that of Bechstein in 1795. We know this because, later, Naumann himself will make it synonymous with the so-called *Stockamsel*.

The *Graue Amsel* [or *Grey ouzel*], a 'grey sort' of *Blackbird,* will continue to crop up in German bird books. For instance, in 1790, Johann Christoph **Stübner** wrote a book about the natural history of the principality of Blankenburg in Saxony in which he says that the *Grauamsel* is slightly smaller than the *Ring ouzel*, travels in pairs, and the male has a dark red spot under its neck which extends to the chest, a feature that distinguishes it from the female which lacks this.[64] He then adds that the *Grey ouzel* is often raised in captivity as if it were a female *Blackbird*, but only a dissection will prove that this bird, with the finding of testicles, is a species different from the *Blackbird*. Stübner thus offers the fundamental observation that the *Grey ouzel* is a male individual that is confused with a female *Blackbird*. This makes it a candidate for Buffon's "autre espèce." Like Naumann, Stübner is convinced that the *Grey ouzel* is a separate species, and that the male *Grey ouzel* exists even if it closely resembles the female *Blackbird*. Yet, if one were to substitute the first-winter male *Blackbird* for Stübner's *Grey ouzel*, then all his findings would make much more sense.

When Johann Andreas **Naumann** wrote the 4th folder of the 1st volume of his *Naturgeschichte der Land- und Wasser-Vögel* in 1797, it was the first and only time that a chapter would be dedicated entirely to the *Stockamsel* as if it were a unique bird species.[65] Naumann supplied us with the most complete details of the bird, though he preferred to use the synonym *Graue amsel* (Grey ouzel) as its principal name. It is thus well worth translating this in its entirety:

Der

Vogelſteller

oder die

K u n ſt

allerley Arten von Vögeln ſowohl ohne als auch auf dem Vogelheerd bequem und in Menge zu fangen

nebſt

den dahin gehörigen Kupfern

und

einer Naturgeſchichte der bekannten und neuentdeckten Vögel

von

Johann Andreas Naumann.

Leipzig,
im Schwickertſchen Verlage. 1789.

Die graue Amsel
The Grey Ouzel
Graudrossel, Braunmerle, Stockamsel

Table XXXI
Figure 64, the male

The *Grey Ouzel* is, on the whole, somewhat shorter and thicker than the *Blackbird*. It is 10 inches long and 16 1/4 inches wide. The color black prevails on the beak, through which at the root of the mandible a beautiful yellow shines forth, the latter of which is also found inside the mouth.

On its head, neck and back, and likewise on the wings, tail and legs it is black-brown or, better yet, smoky black. The throat is stippled with white; the breast is rusty brown and sprinkled with dark, almost round spots; but below the breast and on the belly it loses the brown and becomes ashen grey.

Its nature and call are identical to that of the *Blackbird*, and because the female looks very similar to the male, it has indeed been considered by bird experts to be the female of the *Blackbird*. I ask who is of this opinion, those who want to take the trouble, to dissect the same, and in it they will find testicles in the males and ovaries in the females, and thereby will probably be convinced of their mistake.

It is a rather poor singer. It migrates with the *Blackbird* during the nights of September, October and November, and returns in March and April, though some remain here in the winter.

It nests in our forests, and builds its nest seldom above a man's height, close to trees, on twigs or branches, or even in brushwood, and lines it with rotten wood. In this nest it lays 4-5 eggs and breeds several times, but they are also found in smaller numbers than the *Song thrush*.

In its feeding habits, it behaves just like the *Blackbird*.

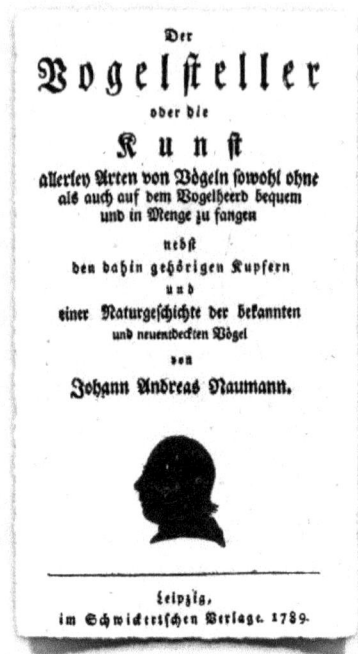

In this book, Naumann completely separates the *Stockamsel* from the *Blackbird* as well as from the *Ring ouzel*, both of which were described under their own heading. He exhibited no hesitancy and was quite adamant of the unique existence of this species, even to the point of daring the dissenters to examine the internal organs as proof of their fallacy. And yet, in sharp contrast to Bechstein, Naumann said that the bird is a poor singer, which leaves us with a feature that is useless for its categorization.

Surprisingly, Naumann even gave the *Stockamsel* an official binomial Latin name (see image below). One will not find this in the body text, but in the index of his text-atlas *Naturgeschichte der Land- und Wasser-Vögel*.

graue Amfel, M. H. Turdus Fuscus. (L.) IV. Tab. XXXI. 64.

This entry tells us that the figure #64 of table XXXI shows an adult male "graue Amsel" in its autumn plumage, and that its Latin name is *Turdus fuscus* as per Linnaeus.

Naumann's figure #64 is the second-ever illustration of a *Stockamsel*. And yet it, too, will meet with a swift demise when some years later, in 1804 to be exact, Naumann himself, in a supplement to this book, will demote the *Graue amsel* to just another form of *Blackbird*. In fact, when he wrote the supplement to his *Naturgeschichte der Land- und Wasser-Vögel* he specifically addressed the *Stockamsel* (or *Graue amsel*) and its illustration in the 1st volume of this book, and he confessed that this figure actually depicted, as one can imagine, an adult female or, even better, a young male *Blackbird*:[66]

> The *Graue Amsel* (Grey ouzel) can be fully crossed out as a separate species, for through my own frequent and reliable observations it appears to be either a young male or an adult female of the *Blackbird*. The female blackbird described on page 143 is an extraordinarily old bird, a time when it is well known that females of several bird species often obtain almost entirely the color of the males. The female *Blackbird* actually possesses all the colors of the bird shown on table XXXI, figure 64, which, however, is a young male in first autumn [i.e., first-winter male] because all the young males of an extremely late brood have this appearance until almost the following spring. By contrast, among those of an early brood one can

Figure of the male *Blackbird* and egg (#63) and the male *Graue Amsel* or *Stockamsel* (#64) in Johann A. Naumann's *Naturgeschichte der Land- und Wasser-Vögel* (1797).

already tell apart the males from the females by the first autumn, and these were those which led me astray so that I, according to the ancient custom of fowlers, considered the *Grey ouzel* as an adult male: therefore, a more accurate description will be necessary.

The whole upper part of the body is dull black, the rearmost wings and the greater coverts of the folded wings are brown; throat white-grey; gullet black with grey-white stripes; the rest of the abdomen dull black, the neck and chest with rusty-brown feather edges, which on the flanks and belly are reduced to white-grey. The russet feather edges form on the crop a type of bowtie and the greater wing feathers are fringed with grey-white; the beak and feet are black brown; the mouth yellow, eyelids brown-yellow. In February I raised a young male (caught in autumn) that perfectly resembled the one described, but he already had a golden yellow beak, black only at its tip, an orange-yellow mouth and bright yellow eyelids. I kept both in a cage and they had sufficient fresh air and sunlight, and in July they were completely like the figure 63 in table XXXI, which illustrates an adult male; they remained black and soon after the moult developed an almost completely black beak, which again was a beautiful orange-yellow in the spring.

Thus, Naumann now considered his figure #64 as a young male *Blackbird* and denied the *Graue amsel*, and consequently the *Stockamsel*, any status as a unique bird species. These new observations by Naumann are of paramount importance since they pave the way for the development of future theories that explain the existence of the 'stockamsel' variety. In fact, Naumann specifically stated that not only is the *Stockamsel* (or *Graue amsel*) a first-winter male *Blackbird*, but also that it is a bird of a late brood; and finally, that the *Stockamsel* will eventually moult into a regular *Blackbird* after one year.

In 1807, Bechstein commented on Naumann's figure of the *Stockamsel* and agreed with Naumann's re-assessment that his figure #64 indeed represented a young male *Blackbird*.[59]

Naumann a. a. O. I. 143. Taf. XXXI. Figur 63. Männchen. Fig. 64. das junge Männchen.

Naumann's re-evaluation of the Stockamsel was documented in 1822 when, along with his son Johann Friedrich, he published the second volume of his *Naturgeschichte der Vögel Deutschlands*.[67] He now gave the synonym of *Stockamsel* to both the *Ring ouzel* (*Turdus torquatus*) and the

Blackbird (*Turdus merula*). When discussing the *Blackbird*, Naumann gave a list of synonyms to which he added that the female and the young bird are also called *Grey ouzel* and *Stockamsel.*

Johann Andreas Naumann

Johann Andreas Naumann (1744-1826) was born into a family of farmers in the village of Ziebigk near Köthen in the district of Anhalt, Germany. In 1760, at the age of 15, he lost his father and had to leave school to help his mother in the management of their farm. He shared with the rest of his family a passion for bird hunting and catching. In addition to hunting birds, he was also interested in their appearance and behavior.

He had three sons, the eldest of which, Johann Friedrich (1780-1857), developed skills in drawing and painting and was soon employed, as early as age 9, by his father to contribute to his ornithological works. He, too, at age 14, had to leave his school in Dessau to help his father on the farm. Their efforts produced the four-volume *Naturgeschichte der Land- und Wasser-vögel* (1795-1803). These farmers did not have the luxury of typing out their manuscript on a computer and emailing off a pdf file to a printing company. No, they did it the hard way. During the winter months, when farming and field hunting activities had ceased, it was then that father and sons organized the notes they had collected during the summer and spent the evenings writing, while during the day they made the paintings and copperplate engravings of their birds, cleaned and polished the copperplates, and printed from them, having also built the presses themselves! The printing of the text and the coloring of the plates were done by others. During the winter they would catch up on the latest ornithological literature and study Natural History, all this supplemented by extensive correspondence with other bird enthusiasts.

The Naumann bird collection, which was initiated by Johann Andreas and continued by Johann Friedrich, is now in the Naumann Museum in Köthen. The Lesser Kestrel (*Falco naumanni*) and the Naumann Thrush (*Turdus naumanni*) were named after Johann Andreas Naumann.

In 1821, Naumann sold his bird collection to Duke Ferdinand of Anhalt-Köthen for 2,000 talers. He was appointed curator of the ducal collection in the newly built Ferdinandsbau in Schloss (Castle) Köthen, where it has been accessible to the public since 1835.

He would continue to accumulate specimens, become one of the founders of the German Ornithologists' Society, and help publish the first official journal of the Society, *Rhea* (two editions). Naumann died on August 15, 1857, and was buried in Prosigk beside his wife Marie Juliane Naumann.

Another depiction of the *Stockamsel* was included in the German-French text-atlas for children *Bilderbuch fur Kinder* written in 1798 by Friedrich Justin **Bertuch**.[68] In it he depicts the "thrushes" that are indigenous to his country, amongst which are the *Blackbird* (*Turdus merula*), the *Ring ouzel* (*Turdus torquatus*), and the so-called *Stockamsel* (*Turdus merula*):

> The *Stockamsel* is brown-black; it has black spots on its neck, breast and belly; and it is probably just a variant of the *Blackbird*. It lives in Germany and its song surpasses that of the *Blackbird*.

In the French text, this bird is given the name *Merle brun-noir*. Bertuch's description would be expanded by Carl Phillip **Funke** in his 1798 book *Ausführlicher Text zu Bertuchs Bilderbuche für Kinder* [Detailed text for Bertuch's Picture Book for children]: [69]

> It appears black-brown, like the female of the common *Blackbird*, provided that the bird which certain naturalists consider the female *Blackbird* is not this variety, as many believe. Some even believe that the song of the *Stockamsel* is more beautiful and stronger than the common one. It also builds its nest on the ground.

The illustration of Bertuch's *Stockamsel* (see next page) resembles much more a bird of the thrush kind than any member of the *Blackbird* family, except maybe the female *Blackbird*. Interestingly, Bertuch gave the *Stockamsel* the Latin name of *Turdus merula* which is identical to that of the common *Blackbird*, thus implying that they belong to the same species.

In Bertuch's next multi-volume book called *Novus Orbis Pictus Juventuti* published in 1807, he added the Hungarian translation of the *Stockamsel* as *A' setétbarna fekete Rigó* (or "dark brown blackbird").[70]

This would be the third illustration of the *Stockamsel* to be printed and, after the retraction by Naumann of his version, this one has survived unscathed from any vociferous criticism to this day. It is hard to know

Fig. 6.

what Bertuch was attempting to depict but, in all honesty, the image has really nothing in common with most descriptions of the *Stockamsel* and resembles far too much a thrush to be considered anything other. It is no wonder that Funke and Bertuch were doubtful of the existence of this variety, and seriously questioned whether it was just the female *Blackbird*.

As to the voice of the *Stockamsel*, there have been divergent assessments. Kramer considered the *Stockamsel* as 'non-singing' and Naumann said that it sang poorly. In 1799, Johannes **Muller** briefly wrote about the *Stockamsel* with a similar appraisal: [71]

> The *Stockamsel* is a type of *Blackbird*; it does not sing as beautifully as the previous [i.e, *Blackbird*], and is brown-black in color, and mottled.

In contrast, Bechstein stated that the *Stockamsel* sang just as well as the *Blackbird*, while Bertuch went even further and said that it sang better. It is quite obvious that these various authors were at odds regarding the singing capabilities of the *Stockamsel*, and randomly assigned diametrically opposite assessments.

CHAPTER EIGHT

Ring ouzel *versus* Blackbird

With the entry on the scene of Bechstein and Naumann, the *Stockamsel* became a common household name in the realm of Central European ornithology. However, the various experts and authors would choose whether to follow Bechstein's writings which equated the *Stockamsel* with either the *Ring ouzel* or the *Blackbird*, or champion Naumann's belief that the *Stockamsel* was its very own unique species. For some curious reason, the *Stockziemer* would not follow the same fate since it would be more often referred to as synonymous with the *Ring ouzel*.

Thus, the rest of the 19th century will experience a divergent understanding of the nature of the *Stockamsel*. There will be books, dictionaries and encyclopedias that mention this bird as a synonym of the *Blackbird*, and others that associate it with the *Ring ouzel*. The following is a table with a list of authors and their interpretation of the *Stockamsel*.

Birds considered synonymous with the *Stockamsel* from the 1700s until today:

Year	Author	Stockziemer	Bergamsel	Ring Ouzel	Blackbird
1756	Kramer[25]		(X)		
1795	Bechstein[57]	X	X	X	X
1795	Goeze[43]		X	X	
1796	*Handbuch* ...[10]	X	X	X	
1797	Bechstein[58]	X		X	X
1797	Bechstein[72]				X
1797	Naumann[a][65]	–	–	–	–
1798	von Heppe[73]	X		X	
1798	Bertuch[68]				X
1799	Muller[71]				X
1800	Bechstein[74]	X		X	
1802	Bechstein[75]	X		X	X
1802	Hopfner[76]		X	X	X
1803	Bechstein[77]		X	X	X
1804	Bechstein[78]	X	X	X	

Year	Author	Stockziemer	Bergamsel	Ring Ouzel	Blackbird
1804	Lippold[79]				X
1806	Orphal[80]			X	
1807	Bechstein[59]	X	X	X	X
1808	Bose[81]	X		X	
1809	Campe[82]	X		X	
1810	Albrecht[83]	X		X	
1812	Bechstein[84]	X	X	X	X
1816	Oken[85]		X		X
1820	Bechstein[86]		X	X	X
1820	Heinsius[87]	X		X	
1822	Winckell[88]	X		X	X
1822	Naumann[67]	X	X	X	X
1823	Brehm[89]				X
1832	Nowak[90]	X		X	
1834	Pierer[91]			X	X
1834	Gloger[92]			X	X
1834	Kaltschmidt[93]	X		X	
1834	Landbeck[94]				X
1834	Hoffmann[95]				X
1834	Fechner[96]				X
1835	Voigt[97]			X	X
1836	Grafe[98]				X
1837	Schilling[99]				X
1837	Oken[100]		X		X
1837	Zander[101]		X	X	X
1838	Weber[102]	X		X	
1839	*Oester...*[103]			X	
1840	Behlen[104]	X		X	
1840	Krünitz[105]	X		X	
1844	Train[106]	X		X	
1845	Schulz[107]				X
1846	Eiselt[108]			X	
1848	Sandmeier[108a]		X		X
1849	Heyse[109]	X		X	
1849	Heck[110]		X		X
1849	Friderich[111]	X	X	X	X
1850	Sandmeier[111a]		X		X
1852	Palliardi[112]				X
1853	Jäckel[113]				X
1853	Speerschneider[114]				X
1854	Willibald[115]			X	
1858	Tschudi[116]	X	X	X	X

Year	Author	Stockziemer	Bergamsel	Ring Ouzel	Blackbird
1860	Wolf[117]	X		X	
1860	Gräßner[118]		X		X
1860	Büchele[119]				X
1861	Mohn[120]			X	(X)
1865	Müller[121]		X		X
1865	de la Fontaine*[122]				X
1865	Homeyer[123]				X
1866	Brehm, AE[124]				X
1868	Sonnenberg[125]	X		X	
1868	Lucas[126]	X	X	X	(X)
1869	Opel[127]	X	X	X	
1869	Schem[128]			X	
1870	Berge[129]	X			X
1873	Russ[130]	X	X	X	X
1875	Meyers[131]				X
1876	Friderich[132]		X	X	X
1877	Giebel[133]				X
1879	*Beschreibung ...*[134]				X
1879	Schmidtlein[135]				X
1882	Reisenthal[136]				X
1882	Fraas[136a]				X
1883	Berghaus[137]			X	
1883	Taubman[138]				X
1888	Thiel's ...[139]				X
1889	Koller[140]				X
1890	Dombrowski[141]			X	X
1890	Liebe[142]				X
1891	Bos[143]	X	X	X	
1891	Jäckel[144]				X
1891	Friderich[145]	X	X	X	X
1891	Ferrant*[146]				X
1891	Olphe-Galliard[147]	X	X	X	(X)
1892	Pechuel/Brehm[148]				X
1892	Floericke[149]				X
1892	Russ[150]	X	X	X	X
1892	Winteler[151]	X		X	
1893	Schmidtlein/Brehm[152]				X
1894	Dalla Torre[153]				X
1895	Löwis[154]		X		X
1897	Meyer[155]				X
1900	Rey[156]			X	X
1902	Krichler[157]			X	

Year	Author	Stockziemer	Bergamsel	Ring Ouzel	Blackbird
1903	Herman[158]				X
1903	Madariaga[159]				X
1904	Ribbeck[160]				X
1904	Bade[161]	(X)	(X)	X	X
1905	Naumann//Blasius[3]		X		X
1906	Knauer[162]				X
1907	Floericke[163]	X		X	X
1907	Schäff[164]		(X)		X
1908	Blanchard[165]	X		X	
1909	Boranić[166]				X
1909	Herzer[167]	X		X	
1911	Fatio[168]				X
1913	Straffen/Brehm[169]			X	X
1915	*Die Gefiederte Welt*[170]			X	
1916	Heyder[171]				X
1919	Tubeuf[172]				X
1930	Sommerfeld[173]				X
1941	Irlweck[174]				X
1943	Morbach[175]				X*
1957	Grimm[176]			X	X
1958	Snow[2]				X
1959	Dornseiff[177]				X
1988	Cramp[178]				X
1998	Desfayes[179]			X	
1999	Stephan[6]				X
2014	Vinicombe[1]				X
2014	Bertau[180]	X			

a, Naumann did not equate the *Stockamsel* with any other species

*, the term *Stackmärel* is equivalent to *Stockamsel*.

(X), elsewhere, the author equates this bird indirectly with the *Stockamsel*.

In 1831, Joseph Leonhard **Hilpert** authored *A Dictionary of the English and German Languages* in which the term *Stockamsel* was attributed to the *Ring ousel*.[181] Conceivably, this would be the first time that the word *Stockamsel* would appear in an English language book, albeit a language dictionary. Hilpert would repeat the same in his *Pocket Dictionary of the English and German Languages* from 1851.

CHAPTER NINE
Brehm

In 1828, Christian L. **Brehm** burst onto the "Stockamsel" scene with the article "Uebersicht der deutschen Vogelarten nach Brehm" that he wrote for the journal *Isis*.[182] In this article he presented a new classification of birds in which he exhibited an unrestrained urge to multiply the number of subspecies of many birds. Brehm placed the *Blackbird* (*Merula nigra*) in its own division and subclassified it into four types, one of which was the *Stockamsel*, albeit without descriptions:

He also classified the *Ring ouzel* into four different types:

In his next book, *Handbuch der Naturgeschichte aller Vogel Deutschlands* [Handbook of the Natural History of every Bird of Germany] published in 1831, Brehm expanded on his four types of *Blackbird* with a couple of paragraphs of description for each.[183] Of the *Stockamsel* he said:

The *Stockamsel.*

Merula truncorum of Brehm.

(*Turdus merula* of Linnaeus)

The elongated beak measures 9''' long from the anterior edge of the nostril to the tip, the crown is scarcely higher than the gently sloping forehead; the nails are long.

It differs from the previous one: 1) by the much flatter head, and 2) the longer toenails; it preferably lives in the deciduous woods of Germany; it likes to build on old tree stumps (*Stöcke,* hence the name) or in brush wood; lays 3 to 5 pale green eggs, sprinkled with reddish and grey; and is similar in other aspects to #1 [i.e., *Fichtenamsel* or *Merula pinetorum*].

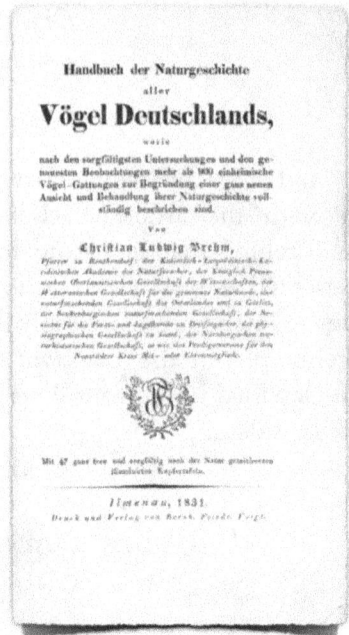

Handbuch der Naturgeschichte
aller
Vögel Deutschlands,

Ilmenau, 1831.

This is the first time that the etymology of the term *Stockamsel* is spelled out clearly and, as suspected, corresponds to "tree-stump ouzel." Of significance, nowhere does the bird's plumage come into play as a special feature of Brehm's *Stockamsel.* This is because he distinguished it purely by its nesting habit.

Brehm would follow this with another book called *Handbuch für den Liebhaber der Stuben-, Haus- und aller der Zähmung werthen Vögel* from 1832.[184] Here, too, he discusses the *Blackbird* and its four types, of which the second type is the *Stockamsel.* Brehm again emphasizes its nesting habits:

> The nest of type nr. 1 (the *Fichtenamsel*) is always in the bushes, higher or lower; that of type nr. 2 (the *Stockamsel*) is on the ground, often on a tree stump, and from there it gets its name *Stockamsel.*

One now understands why Brehm gave the *Stockamsel* the binomial Latin name of *Merula truncorum* which means "Blackbird of the tree trunks." It is clear that Brehm only relied on this bird's nesting site as its most distinguishing feature and made no mention of any unique plumage characteristic. This is plainly evident in his specimen of *Merula nigra truncorum* captured in 1845 and preserved to this very day in the American Museum of Natural History in New York (see pages 57-58 and 60).

Brehm's original specimens of a pair of *Merula nigra truncorum* caught on March 3, 1845, in Renthendorf, Germany (male on left, female on right). Photograph courtesy of Paul Sweet, American Museum of Natural History.

58

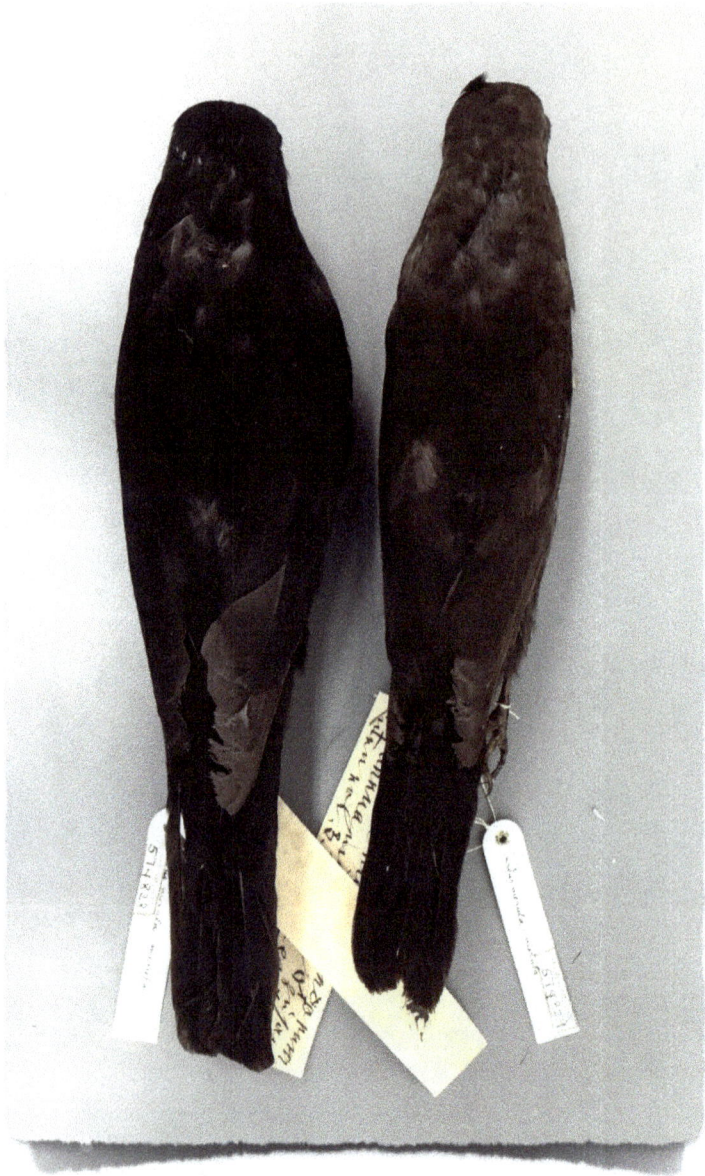

Brehm's original specimens of a male (left) and female (right) *Merula nigra truncorum* caught on March 3, 1845, in Renthendorf, Germany. Photograph courtesy of Paul Sweet, American Museum of Natural History.

Further on, Brehm added:

> I have not yet encountered a male that has worn the plumage of the female, though Bechstein speaks of such.

Brehm goes countercurrent by denying that the male *Stockamsel* has a female-like plumage. This would explain why Brehm's specimen from 1845 resembles so much that of an ordinary *Blackbird*, and why his *Stockamsel* is essentially defined by its nesting site.

Following in Brehm's footsteps, other ornithologists would define the *Stockamsel* only by its nesting habits. For instance, in 1834 Constantin Lambert **Gloger** wrote in his *Vollständiges Handbuch der Naturgeschichte der Vögel Europa's mit besonderer Rücksicht auf Deutschland* [Complete Guide to the Natural History of the Birds of Europe, with special reference to Germany] that the *Blackbird*, to which he gave the synonym of *Stockamsel*, made its nest "in a wide hole of a tree stump or trunk, in firewood, or even in a rock crevice."[92] Friedrich Siegmund **Voigt** was more to the point when in his book *Lehrbuch der Zoologie* (1834) he underscored the difference between the *Blackbird* and the *Stockamsel* based on their nesting habits:

> The expert distinguishes between the *Blackbird* or *Fichtenamsel* and the *Stockamsel*; the latter will build its nest on old tree stumps, the former in the bushes. [97]

Brehm's authoritative words did not dissuade others from concentrating on the plumage as the defining feature of the *Stockamsel*, such as the philosopher and psychophysicist Gustav **Fechner** who, as editor of the 1834 book *Das Hauslexicon: Vollständiges Handbuch praktischer Lebenskenntnisse für alle Stände* [The Home Dictionary: Complete manual of practical life skills for all levels] wrote that the *Blackbird* often retains the color of the juvenile and develops a whitish throat when in captivity, and in such instances it is called a *Stockamsel*.[96] This author also explains the reason for the two alternative terms of *Bergamsel* and *Stockamsel* – the former represents the female *Blackbird* with its distinctly different plumage compared to the male, and the latter is attributed to the harvesting of the male bird in its youth directly from the nest and then raising it in captivity with its unexpected plumage development. This is probably the first time that an author has discussed these two alternate names of the *Blackbird* and has distinguished the two with clarity.

Brehm's *Merula truncorum* from 1845: male (left) and female (right).
Notice the gently sloping forehead of the two birds. The male shows
a hint of pale feather edging on the flank and abdomen.
Photograph courtesy of Matthew Shanley, Staff Photographer, American Museum of
Natural History.

Christian Ludwig Brehm

Christian Ludwig Brehm (January 24, 1787 – June 23, 1864) was a German pastor and ornithologist. He was the father of Alfred E. Brehm, the co-author of the popular book *Brehms Tierleben*. C. L. Brehm was born in 1787, near Gotha, Germany, and studied at the University of Jena. In 1813 he became the minister at Renthendorf, a village sixty miles south of Leipzig, where he remained until his death. His extensive writings included *Beitrage zur Vogelkunde* (1820–22), which described 104 species of German birds in minute detail, and *Handbuch der Naturgeschichte aller Vogel Deutschlands* (1831).

Brehm made a name for himself as an ornithologist by his many publications and an extensive collection of stuffed birds. This collection, held in the parsonage and consisting of over 9,000 dead birds, offered a rare snapshot of the 19th century world of European birds. Brehm offered these to the Berlin Zoological Museum, but the sale fell through. After his death they remained in the attic of his house, where the ornithologist Otto Kleinschmidt discovered them some years later. Kleinschmidt persuaded Lord Rothschild to buy them, and the collection reached his Natural History Museum at Tring located north of London, England. In the 1930s, most of Brehm's bird collection would eventually make its way across the Atlantic Ocean and find a new home in the American Museum of Natural History in New York. And it was here that an example of Brehm's *Merula truncorum* (i.e., *Stockamsel*) was found (see page 57-58 & 60).

Illustration depicting a portrait of German ornithologist Christian Ludwig Brehm sitting in his study, surrounded by books and heaps of dead birds (painting by Carl Friedrich Werner) © Victoria and Albert Museum, London.

Many other scholars could not decide whether the *Stockamsel* was synonymous with the *Ring ouzel* or the *Blackbird* and would thus give the name to both. Still others were quick to dismiss the existence of the *Stockamsel*, such as Christian Ludwig **Landbeck** who in 1834 wrote the *Systematische Aufzählung der Vögel Würtembergs*, a detailed account of the birds of Wurtemberg, a locality in southwestern Germany.[94] He included the *Stockamsel* or *Merula truncorum* of Brehm (*Turdus merula* of Linnaeus) and said that it is "common almost everywhere." However, at the end of the book, under the section "Corrections," Landbeck says that "*Stockamsel, Merula truncorum* Brehm" should be substituted by "*Kohlamsel, Merula atra*, Landbeck" (i.e., the common *Blackbird*). How quickly Landbeck rejected the *Stockamsel* and replaced it with the *Blackbird*, even within his same book!

Another theory that gained steam was that the *Stockamsel* corresponded to only those *Blackbirds* raised in captivity. In fact, when Lorenz **Oken** wrote the multi-volume *Allgemeine Naturgeschichte für alle Stande* in 1837, he said that female *Blackbirds* and those males that are raised in the cage are called *Stockamsel* or *Bergamsel*.[100] Oken astutely realized that the female *Blackbird* in the wild and the males in captivity were the only ones called *Stockamsel*. This notion was not new since, as **Tschudi** and **Winckell** pointed out in their *Handbuch fur Jager, Jagdberechtigte und Jagdliebhaber* from 1858, Bechstein had already pointed out that the description of the *Stockamsel* corresponded to young male *Blackbirds* raised in captivity.[116]

There were then those authors who agreed to dismiss the notion of the *Stockamsel* as a distinct species in toto and believed that this bird was just a form of juvenile male *Blackbird*. For instance, Samuel **Schilling**, who in 1837 published the *Ausfürliche Naturgeschichte des Thier-, Pflanzen- und Mineralreichs*, upon describing the juvenile male *Blackbird* said that some of the young males caught in the autumn, after the first moult, carry all the colors of the female, and that these are commonly called *Stockamsel* or *Graue Amsel* and considered unjustly as a distinct type.[99] Schilling was very perceptive in noting a significant feature of the *Stockamsel*, that is, the dull black plumage of its upper parts, rendering them quite distinct from the browner female. This is more consonant with future depictions of the first-winter male *Blackbirds* and the *Stockamsel* itself. In a similar fashion, William **MacGillivray**, who wrote the second volume of *A History of British Birds* in 1839,[185] made the following remark about the post-juvenile male in the chapter on the *Blackbird*:

> After the first moult, which commences in September, and is completed by the end of November, the plumage of the males is in some almost uniformly brownish-black, while in others the fore-neck, and especially the breast, are more or less lunulated with light-brown and grey.

This would be the forerunner of observations that would introduce the concept of "retarded" plumage of yearling *Blackbirds*, with the latter becoming one of the best possible explanations for the occurrence of the *Stockamsel* variant (see page 85).

And finally, there was Albin **Heinrich** who in 1856 wrote the *Mährens und k. k. Schlesiens Fische, Reptilien und Vögel* in which he combined two of Brehm's four types of common *Blackbird* into one.[186] Thus, the *Fichtenamsel* (Merula pinetorum) and the *Stockamsel* (Merula truncorum) became the hybrid *Merula pinetorum & truncorum*. In other words, Heinrich considered these two birds identical and worthy of only a single designation.

Until now, the *Stockamsel* had been described only in Germany and Austria. The Polish connection would only appear in 1845 when Johann H. **Schulz** wrote his book *Fauna Marchica* on the vertebrate fauna of the March (or State of Brandenburg), located at the time in Prussia and in present-day eastern Germany and western Poland.[107] He included the *Stockamsel* among the trivial names of the *Blackbird*. We finally have a justification, albeit oblique, for the "Poland" reference by Vinicombe et al (2014) (see page 2).

In 1845, C. L. **Brehm** wrote in the journal *Isis* about his discoveries during an 1842 expedition along the Rhine. After shooting a pair of *Merula truncorum*, he commented on a new feature of theirs by saying that they are distinguished by their long beak.[187] This would be reiterated in 1855 when Brehm wrote his book *Der Vollständige Vogelfang* wherein the *Blackbird* was now divided into five types.[188] Here Brehm only stresses the fact that the *Stockamsel* has a longer beak compared to that of the typical *Blackbird* (i.e., *Fichtenamsel* or *Merula pinetorum*), even though in 1831 he initially gave them both the same beak length of 9''' (9 lines).

> Note: The symbol (''') was used to designate a 'line,' an obsolete unit of measure that usually equaled 1/12 of an inch (or sometimes 1/10).

The 5 types of Blackbird in *Der Vollständige Vogelfang* by Brehm (1855).

The confusion surrounding the *Stockamsel* began to take a toll on some authors. For instance, Johann Georg **Heck,** who wrote the *Bilder-Atlas zum Conversations-Lexicon* (1848), said that that the *Stockamsel* was equivalent to the *Blackbird* in general, while the female and young *Blackbird* corresponded only to the *Bergamsel, Grey ouzel* and the *Schwarzdrossel*.[110] Heck was proceeding contrary to all those who had primarily equated the *Stockamsel* with the female and the young *Blackbird*.

Apart from the straightforward etymological clarification of the term "Stockamsel" by Brehm, this period did not produce any major advancement in the understanding of the nature of the *Stockamsel*. Meanwhile, Brehm 'the splitter' could not refrain from creating a new subspecies of *Blackbird* called *Merula truncorum* to represent the *Stockamsel* itself, characterized first by its nesting habit and later by its longer beak. But, as we saw, Heinrich quietly equated the *Blackbird* with the *Stockamsel* and combined them into one single species, the *Merula pinetorum & truncorum*.

CHAPTER TEN

The Missing Link

Up till now, the story of the *Stockamsel* and that of the French *Merle brun* had been moving along parallel tracks without anyone attempting to associate the two. However, on the horizon came forth an Italian naturalist who, though scorned by his contemporaries, would set these birds on a collision course.

In 1858, Gaetano **Perini** wrote the monograph "Degli Uccelli Veronesi" in *Memorie dell'Accademia d'Agricoltura, Commercio ed Arti di Verona*.[189] After describing the *Blackbird* and the *Ring ouzel*, he included the account of a variant of the *Blackbird*:

> Before moving on to other types of birds, we must mention one that differs from our common *Blackbird*, and of which, were it not for our method to employ the highest degree of circumspection, we would be almost inclined to establish as a new species which, by way of our friendship, we will consecrate to the memory of one of the best of our departed friends, that is, Mr. Luigi Menegazzi, who combined the infinite qualities of his soul to the noble love for all things nature, and call it *Turdus menegazzianus*.

> Compared to the *Turdus merula* (Linnaeus), this bird constantly lives in the high hills or mountains, and we have never encountered it on the plains, even though it is occasionally found there by some.

> The dominant color of the male, even in adult plumage, approaches that of the female *Turdus merula* with which it is often confused by our hunters even though the color of the feathers is always darker than the female's.

> The female is also of a uniform shade, in all its plumages, more intense than that of the female *Turdus merula*, and it is very similar to the male who, in the adult nuptial plumage, is never of a deep black such as our *Blackbird*, but always retains a general color of the feathers that is not uniformly black but tending to the brown-olive and ash grey.

> The slender distal margin of the feathers of the throat is constantly of a grey color, as is that of the breast feathers, though less obvious than those of the throat.

> The beak is never yellow-orange, but always horn-brown, except in the spring when it shows a yellowish color that turns horn-brown again in autumn; whereas in the *Turdus merula* this is yellow-orange throughout its entire life.

The color of the eyelid is similar to that of the beak and overall this bird is just a little smaller than the *Blackbird*.

It seldom sings, and when it does it sings only short and interrupted phrases, much less pleasing and harmonious, even though its voice is stronger and louder than the former.

We cannot say anything regarding its food, its nest, or its eggs. A friend of ours, however, after removing them from a nest in the township of Chiesanuova, trained three of these birds, and assures us that they had feathers so pale that he thought they were thrushes (*Turdus musicus*).

Both Bechstein and Naumann mention a variant of *Turdus merula*: the former tries to convince us that it sings well; the latter that it sings worse than the *Turdus merula*.

We present these brief notes to the Ornithologists so that with their investigations and comparative observations they can shed light on a matter which we are unable to perform due to poverty of the former and lack of the latter.

From the description of the plumage and beak, one can confidently conclude that Perini is talking about the *Stockamsel*. And if that is not sufficient proof, he offers confirmation by citing Bechstein and Naumann and their opposite assessments of the *Stockamsel*'s singing capabilities.

The pharmacist Luigi **Menegazzi** (1793-1854) was a passionate amateur entomologist and malacologist as well as a mentor to young Italian naturalists. Born in the small village of Stroppare in the province of Padova, he lived most of his life in Verona, with its surrounding nature providing most of the material for his research. He authored the monograph *Malacologia Veronese* (1855).

However, criticisms were soon to pour in. In 1872, Count Tommaso **Salvadori** wrote in his *Fauna d'Italia* that he believed that Perini's *Turdus menegazzianus* appeared to be individuals of the common *Blackbird* that were not fully adult.[190] In 1902, Count Ettore **Arrigoni degli Oddi** would repeat this opinion in his *Atlante Ornitologico* when he said that the *Turdus menegazzianus* of Perini was nothing more than an incompletely adult male *Blackbird*. He also said that it was of a blackish hue with pale margins of its feathers and had a dark beak; that they nest on the ground and are commonly called by bird catchers as "Merli terragnoli" [Ground Blackbirds] or "Merli di passo" [Migrant Blackbirds].[36] Likewise, in 1903, Pietro **Pavesi** wrote that the faded colors of the underparts and especially the white color with black spots of the throat of Perini's *Turdus menegazzianus* reflected the plumage of a young common *Blackbird* – not of an autonomous species – which indeed has certain thrush-like features that are also observed in the adult female.[191] And so, the *Turdus menegazzianus*, the Italian version of the *Stockamsel*, was ultimately reined in and regarded only as a young (i.e., first-winter) male *Blackbird*.

Salvadori also made the significant observation that such birds had already been described by **Degland & Gerbe** in 1867 who had said that they resembled the *Merle brun* depicted by **Vieillot** in plate 136 of his *Ornithologie Francaise*.[192, 193] Likewise, in 1904, Arrigoni degli Oddi followed suit when he gave the synonym *Merle brun* of Vieillot (*Ornithologie Française*, figure 136) to the *Turdus menegazzianus* of Perini.[194]

One conclusion we can make from all this – besides the plethora of Italian Counts dabbling in ornithology at the time – is that the Italian *Turdus menegazzianus* appears to be the missing link between the Austrian-German *Stockamsel* and the *Merle brun* of the French!

CHAPTER ELEVEN

The Voices of Dissent

By the mid 1800s, serious doubts of the very existence of the *Stockamsel* had begun to creep into the ornithological literature, and the impetus to eliminate it altogether strengthened. The best way to rid themselves of this non-entity was for ornithologists to reduce it to a mere synonym of either the common *Blackbird* or the *Ring ouzel*. One such attempt was by Carl **Mohn** who, in contrast to Brehm the splitter, would be a lumper and attempt to drastically simplify the nomenclature of the blackbirds — although to an extreme. In 1861, Mohn compared one species of *Blackbird* with a matching species of *Ring ouzel* and placed them side-by-side with excerpts of Brehm's descriptions.[120] His pairing of species with similar characteristics was as follows:

Nordische Ringamsel . . . . Stockamsel
Bergringamsel . . . . Hochköpfige Amsel
Gelbschnablige Ringamsel . . . . . Fichtenamsel
Alpenringamsel . . . . Krainische Amsel

He then stated that the *northern Ring ouzel* before the age of 6 to 7 years was probably an ordinary *Stockamsel* which resided in conifer forests in Scandinavia. In Germany, it lived in hardwood forests and built its nest in brush wood. Peculiar only to the genus of the *Stockamsel* and the *northern Ring ouzel* is the fact that they preferred to make their nest closer to the ground. Basically, Mohn believed that the *Stockamsel* would grow up to become a *northern Ring ouzel* — quite a new concept that no one else had ever dared to advance before. Mohn then concluded that each pair of *Blackbird* and *Ring ouzel* corresponded to only one species, and that any difference between them could be ascribed to age. He finally stated that the so-called *Ring ouzels* are only *Blackbirds* which, because of advanced age, acquire a wavy white half-ring on the breast. In conclusion, according to Mohn, the four types of Brehm's *Ring ouzels* are nothing more than older individuals of the corresponding four types of *Blackbirds*. A surprising end to an interesting investigation on the part of Carl Mohn!

It was in 1865 that the two German brothers Adolf and Karl **Müller** wrote a small book titled *Charakterzeichnungen der Vorzüglichsten Deutschen Singvögel.* [Characterizations of the principal German songbirds].[121] Within it, Adolf Müller wrote a chapter in which he remarked how he could not account for the adoption of the *Stockamsel* as a distinct subspecies of

Die Schwarzamsel und ihr Nest.

Illustration in Müller's book that depicts a *Blackbird* and its nest on a tree stump; in other words, a *Stockamsel*.

Die Schwarzamsel und ihr Nest. Nach A. & K. Müller's „Charatterzeichnungen".

Another representation of the *Blackbird* and its nest built among the roots of a tree stump, thus featuring one of the main characteristics of the *Stockamsel*. Notice the menacing fox in the nearby bushes.

the blackbird. He stated that only when it can be confirmed that certain *Blackbirds* nest exclusively in tree stumps, and that this instinct is always propagated from one generation to the next — while other blackbirds never choose this form of nesting — and that such birds have a distinctly different body size, design and special singing talent, only then would they be entitled to form a subspecies "Stockamsel." He said that the habit of nesting on stumps may be a local tendency of *Blackbirds* residing in woodlands where abandoned wood stumps with coppice and knotty growth are common, this being more frequent in the mountains than in the plains. In such localities the bird will often nest upon such knotty and protected tree trunks and stumps. The Müller brothers thus made a strong case for not separating the *Stockamsel* from the *Blackbird* and attributed their presumed distinctive characteristics only to the logistics of the environment in which they are found and where they breed.

A harsher dismissal of the *Stockamsel* would be made in 1865 by Alphonse **de la Fontaine**, author of *Faune du Pays de Luxembourg*.[122] Upon discussing the *Blackbird*, he said:

> In our country, some of the species are migratory, others are sedentary; in the north all the individuals are migratory. These *Blackbirds*, which come through here in October and March, are different from ours, according to the bird-catchers, due to the brightness of their colors, and for this reason they claim that they belong to a separate species. Other people also recognize as a distinct species those *Blackbirds* with a brown or black beak and those that nest in low and dense bushes (*Stackmiérel*). These distinctions are only specious, and they are explained, at least in part, by climacteric [or climatic] influences and by the age of individuals. It would therefore be pointless to refute these beliefs; the error is obvious.

"The error is obvious"! Yet, de la Fontaine does not dignify the reader with any details to support his forceful assertion. There are, however, several interesting facts in this book. When describing the *Blackbird*, de la Fontaine said: "In captivity, it sometimes retains its original color." This observation has been a key element in defining the *Stockamsel* since the times of Bechstein and would eventually be interpreted as a form of "delayed maturation" of the plumage. Furthermore, amongst the synonyms for the *Blackbird*, de la Fontaine does introduce us to a new term: the *Stackmiérel*. This is a word etymologically equivalent to the *Stockamsel*, and he specifies that it is used in the German quarters of Luxembourg.

The aforementioned book by the Müller brothers was reviewed in 1865 by Alexander von **Homeyer** in an article called "Deutsche Singvögel nach A. und C. Müller" for the journal *Magazin für die Literatur des Auslandes*.[123] Homeyer ended by saying:

> and the affair with the *Stockamsel* is considered to be over and forgotten.

And so, one century after its first description, the ornithological world had lost its infatuation with the *Stockamsel* and consequently deemed it non-existent as a unique species.

CHAPTER TWELVE

The Stockamsel Plows Ahead

But the affair did not end. The *Stockamsel* continued to appear in ornithological works as enthusiastically as before. When Alfred E. **Brehm** authored the third volume of his enormously popular book *Illustrirtes Thierleben* [The Illustrated Life of Animals] in 1866, he assigned the synonym of *Stockamsel* to the *Blackbird*.[124] Brehm then introduced the unusual synonym of *Stickziemer* [Stick thrush] to the *Ring ouzel*. This is not a misprint, since this term is repeated in the book's index. Yet this term is immediately substituted with *Stockziemer* by the second edition and will not appear in any other ornithological book. Brehm introduced yet another short-lived synonym for the *Ring ouzel*, the *Stabziemer* [Stick thrush]. This is the first time this term is seen in print, but it did not have much following and was only mentioned in a few German dictionaries. The *Stabziemer*, in fact, is another variation on the *Stockziemer-Stickziemer* series of names.

Brehm's encyclopedic book would be reissued in many editions well into the 20th century. The *Stockamsel* would appear in all editions, but by the last edition of 1913 the term would be applied not only to the *Blackbird* but inexplicably also to the *Ring ouzel*.

Alfred Brehm

Alfred Edmund Brehm (1829-1884) was born on February 2, 1829, in Unterrenthendorf, now called Renthendorf, a small Thuringian village near Neustadt, on the Orla river. He was a German zoologist, as well as a natural history illustrator and writer, the son of the minister and famous ornithologist Christian Ludwig Brehm and his second wife Bertha. His father's research gave Brehm an interest in zoology, but he first began studying to become an architect. His studies, however, were interrupted when Johann Wilhelm von Müller, a famous and wealthy traveller as well as a well-known ornithologist, asked

Brehm to accompany him on an African expedition. Brehm joined the expedition on May 31, 1847, as secretary and assistant to von Müller. The expedition lasted five years and took him to Egypt, Nubia, Sudan, and the Sinai Peninsula; the discoveries made were so important that, at the age of only 20, he was made a member of the German Academy of Natural Scientists Leopoldina.

After his return in 1853, Alfred began to study natural sciences at the University of Jena and graduated in 1855. As is generally the case with extensive travellers, he had no taste for a sedentary career, so in 1856 he went on a two-year journey to Spain with his brother Reinhold. Afterwards he settled down in Leipzig as a freelance writer and wrote many scientific articles and travelogues for various magazines. Apart from this, he undertook an expedition to Norway and Lapland in 1860. In May 1861, Brehm married his cousin Mathilde Reiz, with whom he had five children. In 1862 he accepted the invitation of Duke Ernst II of Saxe-Coburg-Gotha to accompany him on a hunting-trip to Bogosland and Abyssinia. His impressions and observations during this expedition were collected in the book *Ergebnisse einer Reise nach Habesch* [Results of a Journey to Habesch] (1863). Afterwards, Brehm travelled to Africa, Scandinavia and Siberia. His essays and expedition reports from the animal world were well received, and because of this he was commissioned by the editor of the Bibliographisches Institut, Herrmann Julius Meyer, to write a large multivolume work on the animal world. The first six volumes of the encyclopedia, published under the title *Illustrirtes Thierleben*, appeared from 1864 to 1869 and met with wide approval.

The second edition, which consisted of ten volumes published from 1876 to 1879, was titled *Brehms Tierleben* [Brehm's Life of Animals] and it made Brehm famous around the world. The second edition was reprinted from 1882 to 1884, and a third edition, published from 1890 to 1893, followed. The work has been translated into various languages, and editions have continued to appear well into the 20th century, especially in the form of abridged, one-volume works.

Brehm's life was filled with writing, scientific expeditions and lecture tours. Despite this, in 1862, he accepted the post of first director of the Zoological Garden of Hamburg and kept this position until 1867. Afterwards, he went to Berlin where he opened an aquarium. He remained with the aquarium until 1874. In the winter of 1883-1884, Brehm planned a lecture tour in the United States. Shortly before his departure, his four children contracted diphtheria. Since he could not afford to break his contract, Brehm, a widower since 1878, went ahead with his tour. At the end of January, he received word of his youngest son's death. After the hardship of this news, Brehm relapsed into malaria, which he had caught in Africa during his expedition days. On May 11, 1884, he travelled back to Berlin. To find peace, he returned in July to his hometown of Renthendorf, where he died on November 11, 1884. Today, the Brehm Memorial Museum is located there.

Other late 19th century authors would mention the *Stockamsel* in their writings and would justify this term either due to the bird's age or based on its nesting habits. As an example of the former explanation, Karl **Russ** wrote that the young *Blackbird* is sometimes referred to as a *Stockamsel*.[130] On the other hand, **Taubman** said regarding the *Blackbird*:

> The bird-catcher distinguishes the *Stockamsel*, which builds its nest lower, from the *Laubamsel*, which builds it higher; the latter are preferred for raising as they are supposed to be quick to learn. [138]

The *Laubamsel* (literally, "leaf ouzel") is apparently another name for the ordinary *Blackbird*. This is an example of how hunters would create their own idioms to characterize the *Blackbird* according to its nesting site, and specifically on how high off the ground its nest is placed:

Erdamsel	= Ground ouzel
Stockamsel	= Tree stump ouzel
Buschamsel	= Bush ouzel
Laubamsel	= Leaf ouzel

Likewise, Otto **Koller** wrote an article discussing the *Blackbird* wherein he said:

> In 1885, I came across a nest in a pile of wood next to a wooded area. Also, I found that it nests more often in rotten stumps of fallen trees barely 1/2 meter above the ground. People diligently search for the young of such nests, since the "Stockamsel" — such are called the *Blackbirds* whose nest is placed on a tree stump — is said to be a distinguished songbird. [140]

Victor **Ferrant** wrote about his observations gathered during the years 1872 to 1891 in the journal *Bulletin de la Société des Naturalistes Luxembourgeois*.[146] Under the entry of the *Blackbird*, Ferrant said:

> Blackbird, *Mierel*, Turdus merula (Linnaeus). On April 14, a nest with five young birds. They were therefore calculated to have been born at the very latest on March 25. Nests of the *Blackbird*, which are situated in a tree hole or in an old hollow tree stump, often consist only of a simple cavity lined with moss or wool (*Stackmierel*). This location of the nest certainly accounts for the name *Stackmierel*.

Stackmierel is, thus, the equivalent term for the *Stockamsel* in the Luxembourg region.

In 1899, William Thomas **Greene** authored the book *British Birds for Cages and Aviaries: A Handbook relating to all British Birds which may be kept in Confinement*.[195] In the chapter on the *Blackbird*, Greene wrote an interesting paragraph which brings to the fore another problem that has plagued those who have attempted to categorize the female *Blackbird*. It also sheds light on a possible hypothesis regarding the origin of the *Stockamsel* and the like.

> While in their nest plumage the young are much more like their mother than their father, and are even more spotted on the breast than she is. Not long since, one was exhibited as a living proof of the cross-pairing of Thrush and Blackbird, as well as of the production of hybrids in a state of nature. The naturalist to whom the supposed prodigy was shown advised patience, and in due course the bird moulted into a very ordinary cock Blackbird.

Greene addresses the issue of trying to categorize a bird before its complete moult into the adult plumage, and how one can be fooled into classifying it as a different species.

And so ends the 19th century, with no major breakthrough in the scientific knowledge or classification of the *Stockamsel* except for a rehash of previous notions. In general, the authors of this period placed a major emphasis upon the location of the bird's nest in order to justify its existence and separate it from the ordinary *Blackbird*.

CHAPTER THIRTEEN

The Stockamsel in the 20ᵗʰ Century

The 20ᵗʰ century would see several interesting developments in the understanding of moults and plumages of birds. Yet, for the *Stockamsel* it began with the usual banishment to the ever longer lists of synonyms for either the *Blackbird* or the *Ring ouzel*.

One of the most comprehensive descriptions of the *Stockamsel* appeared in volume one of *Naumann's Naturgeschichte der Vögel Mitteleuropas* [Naumann's Natural History of the Birds of Middle Europe] (1905).[3] This was a revised and enlarged edition of Naumann's previous work *Naturgeschichte der Vögel Deutschlands* from 1822.[67] Regarding the *Stockamsel*, the book proceeded to seemingly give one of the most detailed accounts of this bird (even though, upon close inspection, it basically repeats unaltered the text from Naumann's book of 1822). And here we are, 150 years after Kramer had introduced us to the *Stockamsel*, and this bird is still flying high. And yet, it would not be long before the *Stockamsel* will fade to barely a footnote in the ornithological literature.

In 1906, Friedrich **Knauer** wrote the article "Zur Amselfrage" in the journal *Die Umschau*.[162] Upon discussing the *Wald-drossel*, he states:

> Among the *Wald-drosseln* [wood-thrushes] exists a group characterized by the fact that the males are clearly distinguished in color from the females, while in other species the males and females closely resemble each other, so much so that sometimes even the bird expert is unable to distinguish between the two sexes. This group includes the *Blackbird*, in which the adult males are deep black and have a bright yellow beak and a yellow eye-rim, while the young birds and females show a black-brown color with white-grey throat and indistinct dark spots. As a result of this eye-catching gender dimorphism, the fowlers have made two different species of the males and females, the more so as it appears that young males, which are called *Stockamsel* or *Grau amsel*, sometimes retain until their second moult the colored plumage of the females similar to that of the juvenile bird.

Knauer confirms that the *Stockamsel* is indeed a first-winter male that retains the juvenile female-like plumage but will lose this by the following moult. Knauer appears to give the appellation of *Stockamsel* to males only.

In 1909, Hugo **Suolahti** wrote *Die Deutschen Vogelnamen*.[196] He mentioned the Luxembourghese *Stackmierel* as a synonym for the *Blackbird*:

Stackmierel = Stockmerle, d. h. Waldamsel
[Stackmierel = Stockmerle, i.e. Forest ouzel]

The newly minted term *Stockmerle* would be the equivalent of *Stockamsel*. Citing Döbel, Suolahti then mentioned the *Stockziemer* as a synonym for the *Ring ouzel*.

In 1911, the Swiss duo Victor **Fatio** and Théophile **Studer** wrote the seventh volume of their *Catalogue des Oiseaux de la Suisse*.[168] The authors included the synonym *Stockamsle* for the *Blackbird*, a local name used in the Central Plateau and Jura Mountains of Switzerland. The term *amsle* is the Swiss equivalent of *amsel*. Other pertinent synonyms given were *Grauamsle* and *Brunamsle*. The value of this book lies in the fact that it is one of the few times (see also Olphe-Galliard[147]) that the term *Stockamsel* appears in an ornithological book written in French.

And in 1930, with an article written by Sommerfeld, the *Stockamsel* story reached a fork in the road. Some authors would continue to mention its name as just a synonym and maybe explore its etymology while others would go further and search for the physiological explanation of this bird's plumage. The following paragraphs will mention some of the more notable examples of the former group, while the next chapter will introduce Sommerfeld's article and those who followed his path of reasoning.

In 1943, Johann **Morbach** wrote the third volume to his book *Vogel der Heimat* [Birds of the Homeland].[175] The terms *Stockamsel* or *Stockziemer* do not appear in the text. However, the *Blackbird* is given the Luxembourgish synonym of *Stackmärel*. "Stack" in Luxembourgish means "tree stump with rootstock," while "märel" means "ouzel." Therefore, the term *Stackmärel* is equivalent to *Stockamsel*. The tables would turn for this specific name when, in 1998, Michel **Desfayes** wrote *A Thesaurus of Bird Names. Etymology of European Lexis through Paradigms*.[179] Among the German synonyms for the *Blackbird*, he included the *Brunamsle*. However, for the *Ring ouzel* he included the synonyms of *Stockamsle* (Central Plateau of Switzerland), *Stockamsel* (Saxony and Palatinate regions of Germany), *Stockmerle*, *Stackmierel*, *Stackmärel* (Luxembourg), and *Stabziemer*. For reasons unknown, the

Luxemborgish names for the *Stockamsel* were now attributed to the *Ring ouzel*.

Even though the famous brothers **Grimm** (of fairy tale lore) did not live long enough to see their monumental dictionary *Deutsches Wörterbuch* finished, the "S" volume would be eventually published in 1957.[176] Under the entry of *Stockamsel*, it cited both the *Blackbird* and the *Ring ouzel* and then briefly defined the *Stockamsel* as:

> the shy forest bird, which as a nesting place readily chooses the rootstocks in a forest clearing and in the undergrowth.

And so, the 20th century literature on the *Stockamsel* began with a bang owing to the in-depth presentation of this bird in the beautiful ornithology book *Naumann's Naturgeschichte der Vögel Mitteleuropas* — although it turns out that it incorporated the very same passage from Naumann's *Naturgeschichte der Vögel Deutschlands* of 1822, itself a rehash of Naumann's own observations hidden in a little-known supplement from 1804. The *Stockamsel* story would then continue along two paths, one of which was littered with brief citations of the term as a mere synonym for either the *Blackbird* or the *Ring ouzel*. The other path would attempt to explore the background physiology within which this avian curiosity existed, and this began with the pivotal study by the German ornithologist Eckhart Sommerfeld who took some crucial assertions by Naumann and others and developed them further.

CHRONOLOGY OF THE STOCKAMSEL

1756 = Kramer is the first to mention the *Stockamsel*.

1789 = Naumann introduces the *Grau amsel* [Grey ouzel].

1795 = Bechstein publishes the 1st image of the *Stockamsel* as a *Ring ouzel*.

1795 = von Schauroth mentions the nesting site of the *Stockamsel*.

1797 = Naumann publishes the 2nd image of the *Stockamsel*.

1798 = Bertuch publishes the 3rd image of the *Stockamsel*.

1800 = Sonnini introduces the *Merle brun* [Brown ouzel].

1804 = Naumann demotes the *Grau amsel*; proposes the late brood theory; observes that *Stockamsels* will moult into all-black *Blackbirds*.

1824 = Dumont proposes the delayed acquisition of the *Blackbird's* adult plumage to explain the *Merle brun*.

1831 = Brehm explains the etymology of the word *Stockamsel*, and gives it the name *Merula truncorum*.

1856 = Perini coins the name *Turdus menegazzianus*.

1865 = Müller brothers debunk the existence of the *Stockamsel*.

1930 = Sommerfeld publishes the 1st photograph of a *Stockamsel*; proposes the hormonal theory.

1958 = Snow presents the 1st English language discussion of the *Stockamsel* and links it to the concept of *Hemmungskleid*.

1988 = Cramp introduces a modern and faithful illustration of the *Stockamsel*.

2016 = One of Brehm's *Merula truncorum* from 1845 resurfaces within the ornithology collection held in the American Museum of Natural History.

CHAPTER FOURTEEN

A Modern Theory

It was in 1930 that Eckhart **Sommerfeld** wrote the article "Gefiederstúdien an Drosseln" [Study of the Plumage of Thrushes] in the journal *Anzeigers Ornithologische Gesellschaft in Bayern.*[173] After examining a total of 60 *Blackbirds*, he described in detail the various age-specific plumages of this bird. Upon discussing its first-year coat, he said that one can easily distinguish the male *Blackbird* in its post-juvenile plumage from the subsequent adult, namely by the brown wings. One can even establish whether this or that individual is a bird of an early or late clutch: in the former case the plumage will be all black after the post-juvenile moult; conversely, in the latter case we are dealing with a so-called *Stockamsel.* Thus, Sommerfeld echoes the notions advanced by Naumann in 1804 and MacGillivray in 1839 that there are at least two types of first-winter male *Blackbirds*: one from an early brood that looks almost like an adult, while the second is from a later brood and has the plumage of the *Stockamsel* type.

Sommerfeld proposed a physiological cause for the variation in plumage appearance of post-juvenile *Blackbirds*. If they are of a later brood, their gonads are at a younger stage than those of an early brood. Thus, upon the post-juvenile moult they appear as *Stockamsels* with discolored feathers that have more or less retained their previous features, since the hormonal blood levels are too low to allow it to reach a fully colored plumage. Proof that the *Stockamsel* is younger than a one-year-old adult *Blackbird* can be found in the cranium. This is still cartilaginous in early November in the *Stockamsel* while its ossification would be almost complete in the adult at this time. Finally, in the *Stockamsel* the yellow-orange staining of the beak occurs later.

In this article, Sommerfeld depicted the first known black & white photograph of a *Stockamsel*, albeit dead (see next page).

Its caption reads:

> *Stockamsel* with discolored feathers on the right and left side of the chest and abdomen, not in the subsequently moulted middle of the chest. The contrast is not as stark as the figure indicates.

In this picture, one sees the chest and abdomen unevenly blotched in white. It does not appear to be a very fortunate image since alternate interpretations of such fading could be advanced, such as a *Blackbird* with partial albinism or a *Pied blackbird*. On the other hand, one does notice distinctive pale edging of the breast and abdominal feathers, a feature that will be stressed in the future as typical of the *Stockamsel*.

Sommerfeld's concept of the *Stockamsel* as representing a post-juvenile, first-winter *Blackbird* that most likely hatched from a late clutch will be reprised by the British ornithologist David Snow some three decades later.

It was in 1958 when David W. **Snow** published the first edition of his book called *A Study of Blackbirds* (a second edition was released in 1988).[2,197] The second chapter dealt with plumages and moults. Snow eloquently described the plumage of the *Blackbird*, both male and female, as well as of the young (see page 2). He included a passage wherein he stated that males in retarded first-year plumage, namely brown-black with a whitish throat, were known as 'Stockamsel.' This was hinted to by earlier authors such as Naumann, Dumont, MacGillivray, and Arrigoni degli Oddi. Incidentally, this may be the first occurrence of the term "Stockamsel" in the English ornithological literature.

The concept of "Hemmungskleid" or "retarded plumage" had been introduced and defined by Erwin **Stresemann** in 1919 when, upon discussing the plumage of the *Black-headed Bunting*, he said that not infrequently males have a "retarded plumage" (i.e., their plumage remains similar to that of the female) and that this was probably associated with the inadequate internal secretions of the sex glands.[198]

In 1980, Sievert **Rohwer** et al coined the phrase "delayed plumage maturation" to describe the delayed acquisition of adult plumage by sexually mature birds.[199] This retention of immature plumage after the post-juvenile moult produces a female-like plumage in yearling males, i.e., intermediate between that of females and the definitive (adult) male plumage. These birds will acquire a definitive color and pattern of plumage after the first potential breeding period. This phenomenon occurs in birds with sexual dimorphism in which the plumages of the adult male and female are very different. This is certainly the case of the *Blackbird* which shows a marked sexual dimorphism with an entirely black plumage in adult males and a brownish one in females. In addition, both bill and eye ring are bright yellow to orange in males, whereas females have brown bills admixed with dull yellow areas in some.

Various terms such as female-like, juvenile-like, immature, and the somewhat reprehensible "subadult," have been used to refer to an individual, usually male, that has not acquired definitive plumage color and pattern by its first potential breeding period. It must be stressed that "delayed plumage maturation" occurs after the post-juvenile moult when a bird ought to show an immediate acquisition of a mature or adult-like plumage. In other words, it concerns those intermediate plumages between juvenile and full adult plumages. Thus, juvenile plumage is not considered part of the "delayed plumage maturation" phenomenon. In regard to the term "subadult," the journal *British Birds* advises against the use of this term, with preference for alternate terms such as "immature," "immature other than juvenile," or "near adult."[200]

It is unclear from the literature whether the "retarded plumage" of Stresemann corresponds to "delayed plumage maturation." It would appear that the two are synonymous, either fully or in partial form. Only rarely has the latter expression been associated with the *Blackbird*.

86

One such instance is the article by **Escalona-Segura** & Peterson (1997) wherein the authors say:

> Published documentation of strikingly delayed plumage maturation is surprisingly scant in the subfamily Turdinae. Three additional species that do show delayed plumage maturation are Turdus Merula (Snow 1958), T. olivater (K. C. Parkes) and T. rufitorques (Skutch 1960). [201]

As many as 15 hypotheses have been advanced to explain "delayed plumage maturation." These include status signaling, female mimicry, crypsis, juvenile mimicry, and moult constraint. The consensus is that "delayed plumage maturation" is part of a strategy to reduce competition from older individuals, often in a reproductive context, during the first year of life. For instance, young males with an immature plumage signal their low or subordinate status to reduce aggression from older males; a dull plumage mimics female coloration and therefore, through deception, reduces male aggression and increases the chance of breeding and of acquiring female-worthy territories; and a juvenile-like plumage mimics non-reproductive juveniles to avoid aggression from the adult males. As for the the dusky bill described in the *Stockamsel*, this could also agree with the *delayed maturation* hypothesis since the development of a showy, bright yellow-orange bill may be costly in yearlings.[202]

In 1988, Stanley **Cramp** edited volume five of the monumental series *Handbook of the Birds of Europe, the Middle East and Africa. The Birds of the Western Palearctic.*[178] Among the *Blackbirds*, this book included an image of the *Stockamsel* (see next page) designated in the caption as:

> 1st adult male fresh (autumn) 'stockamsel' variety.

The *Stockamsel* appears in the above figure as blackbird #4. No mention of this variety appears in the body text. The depicted *Stockamsel*, when compared to the regular juvenile, differs by its pale underparts which have a reticulated pattern created by the prominent grey edging of the feathers on its chest and abdomen, and by the white streaks on its chin and throat. This pattern is reminiscent not only of the adult female *Blackbird* but even more so of the *Ring ouzel*. It appears quite obvious why both types of bird species have been considered synonymous with the *Stockamsel*.

And finally, Keith **Vinicombe** et al wrote that first-winter male *Blackbirds* of the so-called 'stockamsel' type (from Germany and Poland) have a dull bill and eye-ring, browner wings, a paler chin, and heavy pale fringing to the underparts feathers.[1]

1: adult male; 2: adult female; 3: juvenile male; 4: 'stockamsel' variety.

Blackbirds (*T. merula*) from the *Handbook of the Birds of Europe, the Middle East and Africa* - vol. 5: "Tyrant Flycatchers to Thrushes" edited by Cramp (1988), Plate 70, 1-4. By permission of Oxford University Press.

Thus, the consensus in the more recent literature is that the 'stockamsel' type of *Blackbird* is a young (first-winter) male *Blackbird* before its first adult moult. It has some features of the female *Blackbird*, such as a brown bill, lack of a bright yellow eye-ring, browner wings, paler feathers of the chin and neck, and grey edging of its underside feathers. Such features appear to be quite similar to the female *Ring ouzel*, in whom the white neck ring is duller and sometimes quite indistinct. This would explain the divergent assignment of the term 'stockamsel' to the *Blackbird* and the *Ring ouzel* over the centuries. What is surprising is that in all those years not one author had put their foot down and pronounced one bird versus the other as the true *Stockamsel*. This could only suggest an understandable insecurity on their part as to which one was the real *Stockamsel*, considering that the *Blackbird* and

Ring ouzel are confusion species themselves and easily mistaken one for the other at certain stages of their life.

First-winter male *Blackbird* with female-like underparts & some juvenile brown wing-feathers; in other words, a 'Stockamsel.' Photograph © by Anthony McGeehan.

Five principal characteristics that define the *Stockamsel* were cited by Vinicombe et al.[1] If we compare these to the ones described in the past literature (see list in footnote), we notice either a variance or a consistency in the perceived characteristics of the *Stockamsel*:

ABDOMEN — heavy pale fringing (as per Vinicombe)
- 1795a   ashen black
- 1795b   dull black, grey or brown feather edges
- 1797a   lighter
- 1797b   ashen grey
- 1798   black spots
- 1803   rusty yellow with wavy black spots
- 1804   dull black with white-grey feather edges
- 1807   dirty white-grey
- 1820   ash grey
- 1837   dull black
- 1858   brighter

Ref.: 1795a: Bechstein[57]; 1795b: von Schauroth[57]; 1797a: Bechstein[58]; 1797b: Naumann[65]; 1798: Bertuch[68]; 1803: Bechstein[77]; 1804: Naumann[66]; 1807: Bechstein[59]; 1820: Bechstein[86]; 1837: Schilling[99]; 1858: Tschudi.[116]

The descriptions of the abdomen ranged from dirty white-grey to ashen black, but some did describe it as dull black with grey-white edges to the feathers, similar to today's description (see the photograph on the next page taken by Irish bird photographer Anthony McGeehan).

CHIN / THROAT — paler (as per Vinicombe)
    1795[a] white-grey with black stripes
    1797[a] whitish with dark brown stripes
    1798   stippled with white
    1803   blackish spots mixed with rusty yellow
    1804   black with grey-white stripes
    1807   whitish
    1820   whitish
    1858   white roan, dark brown stripes

This second feature is also quite evident in Cramp's image. The throat has white admixed with linearly arranged small dark spots and is similar to that seen in the female *Blackbird*. This characteristic has been consistently documented by the ornithologists over the years.

WINGS — browner (as per Vinicombe)
    1795[b] dull black, grey or brown feather edges
    1797[b] black-brown or smoky black
    1804   fringed with grey-white; brown wing coverts

Regarding the wings, we can draw the conclusion that they undergo partial moulting with new black feathers and retained older juvenile brown ones, each with varying degrees of grey edging. The presence or absence of prominent grey-white fringes to the wing feathers would determine if a particular example of *Stockamsel* should be considered a *Ring ouzel* or a *Blackbird*, respectively. For instance, Naumann in 1797 was describing a *Blackbird* with "black-brown" wings, whereas Naumann in 1804 may have been portraying a *Ring ouzel* with its wings "fringed with white-grey."

BEAK — dull (as per Vinicombe)
    1795[a] orange-red and brown at the root
    1795[b] only one bright yellow
    1797[a] half black-half yellow
    1797[b] black, with yellow at its root
    1804   black-brown
    1837   half yellow
    1858   half black, half yellow

The color of the beak has varied immensely among the various descriptions, ranging from yellow to orange to "dull." The latter term implies a brown bill with or without shades of faint yellow focally present. It would appear that each case is different based upon the amount of yellowing of the beak, but the tendency for a darker beak appears to be a constant identifying feature.

First-winter male *Blackbird* with female-like underparts & mottled brown-dull yellow beak; in other words, a classic 'Stockamsel.' Photograph © by Anthony McGeehan.

EYE-RING — dull (as per Vinicombe)
  1795[a] white-yellow
  1795[b] not yellow, almost white
  1804   brown-yellow

The constant theme here is that the eye-ring is distinctly not bright yellow, but a duller yellow or whitish ring.

All these findings basically confirm that the *Stockamsel* is a first-winter male *Blackbird* in which certain non-adult features are accentuated, thus supporting the "retarded plumage maturation" theory.

The reason that some authors had considered the *Stockamsel* as a variant of the *Ring ouzel* is understandable since the descriptions of the *Stockamsel* contain some characteristics of the first-winter male *Ring ouzel*. Thus, there is some support to the idea that the *Stockamsel* could also represent a young *Ring ouzel*, especially since one of the *Stockamsel's* defining features is the visible pale fringing of the underpart feathers.

Descriptions of the adult female *Ring ouzel* also contain several key elements of the plumage of the *Stockamsel*, and it makes sense that the adult female *Ring ouzel* may have been considered by some as a *Stockamsel* variety of ouzel.

In summary, based upon a spectrum of changes in their plumage, there are several variations of *Stockamsel* described in the literature. In order of frequency and likelihood, these would correspond to the following:

(1) A first-winter male *Blackbird* with prominent pale feather edging on the chest and abdomen, presence of black and brown wings, a brown or mottled yellow-brown beak, and absence of prominent pale fringes to the wing feathers;

(2) A female *Blackbird*, especially an older adult which tends to acquire a blacker plumage with advanced age;

(3) A young male or an adult female *Ring ouzel* which have similar features as the first-winter male *Blackbird* with the addition of a prominent grey fringe to its wing feathers;

(4) An ordinary *Blackbird* with no distinguishing plumage feature other than the fact that it nests on tree stumps, preferably in forests (see Brehm).

Details of a 'Stockamsel'
photography by Anthony McGeehan.

Beak: color only partially yellow-orange, typical of a male in post-juvenile plumage in early spring.

Wings: well defined boundary between black (moulted) and brown (non-moulted) feathers.

Chest & Abdomen: noticeable grey feather edging (occasionally with a rufous tinge) on the underpart's feathers.

CHAPTER FIFTEEN

Concluding Remarks

The *Stockamsel* is a bird that has struggled to become its own distinct species. It has often been associated or identified with two birds, the *Common blackbird* (*Turdus merula*) and the *Ring ouzel* (*Turdus torquatus*). It was originally paired with other bird types whose names have not withstood the test of time, such as the shadowy *Stockziemer*, the mythical *Merula montana*, the lofty *Bergamsel*, the furtive *Rock ouzel*, and even the ancient *Merula fusca* of Aristotle.[203] The latter has been associated with the *Ring ouzel*, the *Solitary sparrow* (today's *Blue rock thrush* or *Monticola solitarius*), and even the indeterminate *Avis venatica* of Belon.[204] Other birds that seemingly coincide with the Austro-German *Stockamsel* are the long-lost *Merle gris* and *Merle brun* of the French, the *Turdus menegazzianus* of the Italians, and the *Graue amsel* of the German ornithologists.

This confusing scenario reflects several problems with the early age of Ornithology, such as the assignment of multiple names to the same bird, the artificial multiplication of bird types due to lack of knowledge of the various moulting plumages of a certain bird species, the prominent dichotomy of plumage between the sexes of the *Blackbird*, and the dilemma with the separation of two similar species comprising birds with overlapping features, such as in the case of Merula (merle / blackbird) *versus* Turdus (grive / thrush).

Such dilemmas were eloquently expressed in 1802 by the ornithologist François **Levaillant** in his book *Histoire Naturelle des Oiseaux d'Afrique*.[205] Levaillant made the point that the difference in behavior between two similar kinds of birds can and should be used as a factor in justifying their categorization as two separate and distinct species. On the other hand, he advises against relying excessively on certain plumage characteristics when splitting bird types because of the variability that can occur within a single species due to age, gender, or moulting phases.

Indeed, those who described the *Stockamsel* may well have fallen into this trap since this bird, after having been initially separated out because of its nesting habits, was thereafter rendered distinct because of its plumage features. But these differences were gradually deflated and attributed only to environmental factors and aberrant plumage variations.

In conclusion, the *Stockamsel* was a short-lived bird "species" that was introduced as a unique type of bird but quickly lost this privilege. It was initially associated with a group of birds, some of which have populated the written history of Ornithology since ancient times. In fact, Aristotle's *Merula fusca* was one of the first birds to be equated with the *Stockamsel*. Thereafter, it became identified with only a small number of more or less well-defined bird types.

Of the two major bird species that have been identified with the *Stockamsel*, the one with the longest running history has been the *Blackbird* (*Turdus merula*), a bird well known since the time of Aristotle. The dichotomy of plumage of the *Blackbird*, which reflects the marked sexual dimorphism between adult males and females, was an issue that beleaguered the field naturalists of the time. The features of the female are so distinct that many have confused them with the speckled plumage of thrushes. It is no wonder that we have the term *Merle-grive* introduced by befuddled bird catchers. The young of the *Blackbird* resemble their mother and have been a source of confusion too. In contrast, the adult male *Blackbird*, due to its distinctive all-black plumage, has never been a contender as the real *Stockamsel*, except by those who relied solely on this bird's nesting habits (see C. L. Brehm).

The *Ring ouzel* (*Turdus torquatus*) is a more recently described bird and has been recognized as a distinct entity since the mid 1500s. The male representative of the *Ring ouzel* creates few problems due to its distinct white neck band. However, it is the female again which can cause confusion. This is because she has a browner coat and her neck band is of a duller and dirtier white which sometimes verges on the indistinct. Like the *Blackbird*, the young *Ring ouzel* resembles its mother. It is especially the *Ring ouzel*'s distinctive grey edging of the feathers on its chest and abdomen that is reminiscent of certain descriptions of the *Stockamsel*. Even though today most ornithologists equate the *Stockamsel* with the *Blackbird*, it is understandable why so many in the past have preferred to relate it to the *Ring ouzel*, and why some were even unable to decide between the two, and thus associated the *Stockamsel* with both.

Another bird akin to the former two is the *Bergamsel* or "Mountain ouzel." Its story began in the early 1500s and continued for a good 300 years. The *Bergamsel* followed the fate of the *Stockamsel* in that it, too, was briefly given the honor of its own formal species but would soon gain anonymity by being considered a synonym of the *Stockamsel* and, later, as just another synonym for either the *Blackbird* or the *Ring ouzel*.

The story of the *Stockamsel* itself is one of the youngest, having been first mentioned only in the mid 1700s. It would briefly be given the privilege of being considered a distinct species, but it would not be long before its very existence would be severely questioned and it would end up as just a synonym for either the *Blackbird* or the *Ring ouzel*, and specifically of the adult female or the young male.

Local variations on the name *Stockamsel*		
Stockamsel	Austria-Germany	(Kramer, 1756)
Stackmierel	Luxembourg	(de la Fontaine, 1865)
Stockamstel	Tyrol	(Dalla Torre, 1894)
Stockmerle		(Suolahti, 1909)
Stockamsle	Switzerland	(Fatio, 1911)
Stackmärel	Luxembourg	(Morbach, 1943)

A review of the literature reveals that, by 1850, the *Stockamsel* had been identified with the *Ring ouzel* slightly more often than with the *Blackbird* (28 citations *vs* 27 citations). This is most likely because many authors relied on Bechstein's description and his pictorial depiction of the *Stockamsel* as a *Ring ouzel*. Also, the dictionaries of the time tended to favor the *Ring ouzel* as synonymous with the *Stockamsel*. Only in the 20th century did the trend reverse itself, with the *Blackbird* as the favored representative of the *Stockamsel* (at least 82 *Blackbird* citations *vs* 51 *Ring ouzel* citations as of 2015). If we eliminate the various entries in lexicons, dictionaries and encyclopedias (which gave preference to the *Ring ouzel*) and include only genuine ornithological works, we see a more emphatic outcome for this data (73 *Blackbird* citations *vs* 35 *Ring ouzel* citations). The *Stockziemer*, on the contrary, has almost always been associated with the *Ring ouzel*.

Many other birds have been included in the list of synonyms for the *Stockamsel* along with the *Blackbird* and/or the *Ring ouzel*. Only in rare instances were the latter two not included, and in those instances the *Stockamsel* would be identified with another bird, this being either the

Meeramsel (four times), *Grey ouzel* (four times) or the *Bergamsel* (two times). The latter three birds have essentially disappeared from the ornithological literature after being absorbed into the categories of either the *Blackbird* or the *Ring ouzel*. Only three times (Kramer in 1756, Naumann in 1797, and Bertuch in 1798) has the *Stockamsel* ever been considered a separate entity, but always in close company with the *Blackbird* and the *Ring ouzel*.[25,65,70]

Regarding the first official illustrations of the *Stockamsel* by Naumann and by Bertuch, these appear thrush-like and at the same time similar to the female *Blackbird*. Moreover, in Naumann's figure the *Stockamsel* is placed on a page together with a male *Blackbird* and its egg. Considering the company that it keeps, it is automatic to assume that this bird should be a *Blackbird*, whether an adult female or a young male. To Naumann's credit, he did rectify this.

Proposed Latin binomial names for the *Stockamsel*	
Turdus fuscus	Naumann (1797)
Turdus merula	Bertuch (1798)
Merula truncorum	Brehm (1828)
Turdus menegazzianus	Perini (1858)

The *Stockamsel* has existed in the world of Ornithology for the past two and a half centuries, mainly within the German ornithological literature, but also in the French literature as the so-called *Merle brun*, *Merle-grive* or *Merle gris*. It briefly appeared in the Italian panorama but was quickly shot down by eminent ornithologists of this country. The relative lack of its mention outside of the central European continent does not necessarily imply that the *Stockamsel* has a particular geographical distribution. After notable German ornithologists had elevated its status to that of its own species, the *Stockamsel's* seemingly heavy presence in Central Europe may be related to the ostensible tendency of authors to repeat a compatriot's assertion without independent confirmation. Today, the *Stockamsel* is occasionally mentioned in the British and German literature, as well as on internet chat groups, to describe a variant of the *Blackbird* which, by many, is interpreted as a first-winter male with delayed or arrested maturation of its plumage.

And as to when and why the *Ring ouzel* fell into disfavor as the candidate bird associated with the *Stockamsel* — nothing is known.

APPENDIX A

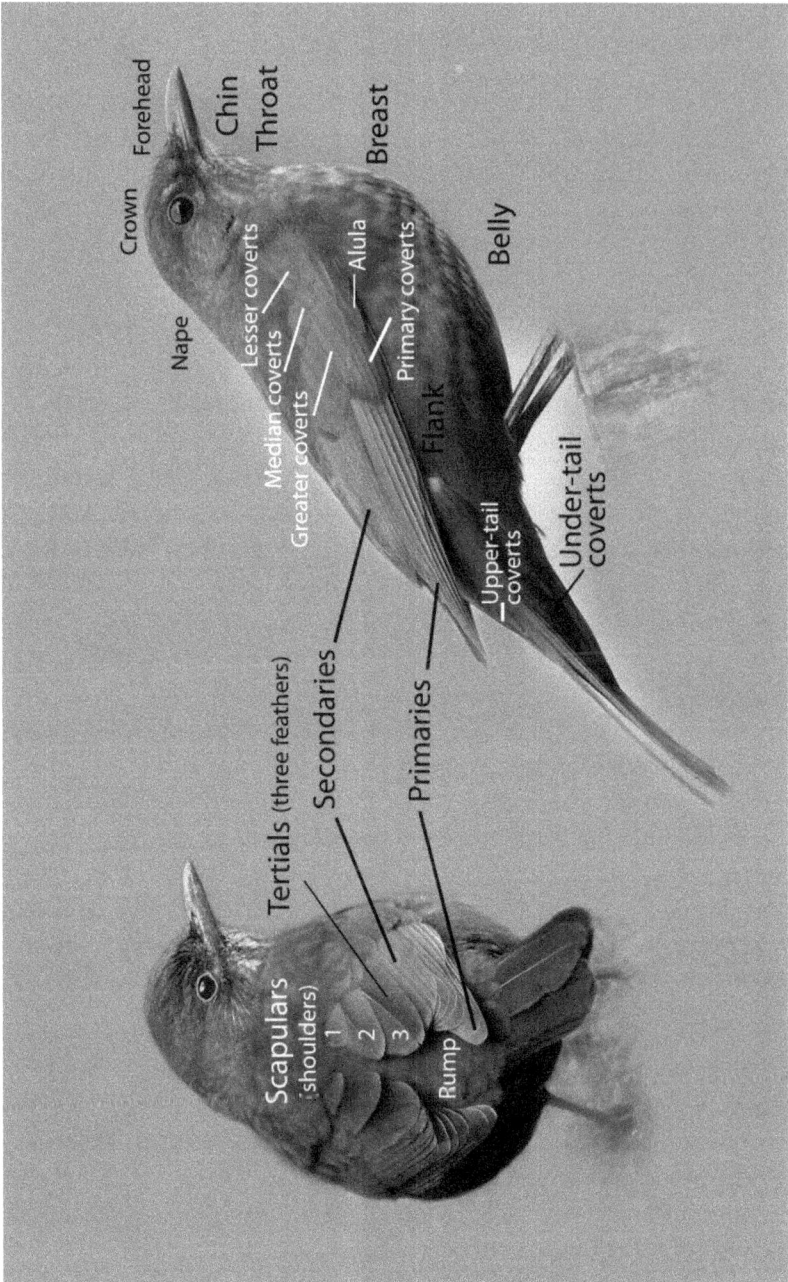

"Topography of a Blackbird" © by Anthony McGeehan

APPENDIX B

Plumages and Moults of the Blackbird

Upon hatching, the *Blackbird* chick is born with a sparse *downy* plumage. After a week, it grows its first feathers (*postnatal moult*) and develops a spotted and streaked *juvenile* plumage. While in the nest, the male and female juveniles are hard to distinguish, but when they are old enough to leave the nest, the males can be distinguished by their darker colors.[2]

After three months, the *Blackbird* moults into a *post-juvenile* (or *first-winter*) plumage which is similar but not identical to the full adult plumage. During this moult, all the feathers are changed except for certain wing feathers (primaries, outer secondaries, alula, primary and greater coverts). This post-juvenile plumage will carry the first-year bird through the winter and into the breeding season until the following autumn. In the male, the fresh new feathers are not as glossy black as those of the full adult type and the breast feathers will have some brown edging. The male's non-moulted juvenile feathers of the wing with their pale brownish color notably contrast with the new black feathers. The female has pale brown to rufous feathers, particularly on the breast. The young females are paler and more ginger than their adult counterpart. The first-year plumage of males can range from an almost full adult black (advanced) to a female-like brown (retarded). It is from the latter end of the spectrum of this wide range of first-year plumage in males that we have the "stockamsel" variant; i.e., young adult males in retarded plumage. The beak and eye-ring of the first-year bird remain dark as that of the juvenile, but in the winter the beak will begin to show a smattering of yellow in the male.[2]

In its second autumn, the *Blackbird* acquires a new plumage through a complete moult wherein it acquires a new and complete set of black feathers. This *full adult* plumage does not change appreciably at subsequent moults. Some females become darker with age. The beak and eye-ring become yellow-orange and remain so, albeit somewhat duller each autumn.[2]

....

REFERENCES

[1] 2014 – Vinicombe, Harris & Tucker: *The Helm Guide to Bird Identification: An In-Depth Look at Confusion Species.*

[2] 1988 – Snow: *A Study of Blackbirds* – 2nd edition

[3] 1905 – Naumann: *Naturgeschichte der Vögel Mitteleuropas* - band I

[4] 1980 – Baillie: "The extent of post-juvenile moult in the blackbird" in *Ringing & Migration*, 3:1, 21-26

[5] 2009 – Khokhlova: "Juvenile moult and spatial behaviour of first-year Blackbirds *Turdus merula* on the northeast edge of the range" in *Avian Ecology and Behaviour*, 15: 1-22.

[6] 1999 – Stephan: *Die Amsel* - 2nd edition

[7] 1902 – Seebohm: *A Monograph of the Turdidae*

[8] 1920 – Witherby: *A Practical Handbook of British Birds*

[9] 1759 – Großkopff: *Neues und Wohl eingerichtetes Forst- Jagd- und Wiedewercks Lexicon*

[10] 1797 – *Handbuch für Praktische Forst- und Jagdkunde* - volume III

[11] 1892 – Dombrowski: *Allgemeine Encyklopaedie der gesammten Forst- und Jagdwissenschaften* - volume VII

[12] 1742 – Zorn: *Petino-Theologie* - volume II

[13] 1555 – Gesner: *Historiae Animalium: qui est de Avium Natura* - liber III

[14] 1772 – *Onomatologiae Forestalis-Piscatorio-Venatoriae Supplementum* - volume I

[15] 1984 – Lockwood: *The Oxford Dictionary of British Bird Names.*

[16] 1483 – *Catholicon Anglicum* (see Herrtage, 1882).

[17] 1760 – Klein: *Verbesserte und vollständigere Historie der Vögel.*

[18] 1750 – Klein: *Historiae Avium Prodromus*

[19] 1505 – Melber & Eichmann: *Vocabularius Predicantium*

[20] 1545 – *Thierbuch* (see Albertus Magnus: *De Animalibus*, 1478)

[21] 1746 – Döbel, HW: *Jäger-Practica*

[22] 1828 – Döbel: *Jäger-Practica* - 5th edition

[23] 1912 – Döbel: *Jäger-Practica* - 6th edition

[24] 1750 – *Allgemeines Haushaltungs Lexicon* - volume II

[25] 1756 – Kramer: *Elenchus Vegetabilium et Animalium Per austriam Inferiorem Observatorum.*

[26] 1896 – Newton: *A Dictionary of Birds*

[27] 1676 – Willughby: *Ornithologiae* - volume II

[28] 1713 – Ray: *Synopsis Methodica Avium & Piscium*

[29] 1746 – Linnaeus: *Fauna Svecica*

[30] 1600 – Aldrovandi: *Ornithologiae - Tomus Alter* - liber XVI

[31] 1555 – Gesner: *Icones Avium Omnium* - 1st edition

[32] 1761 – Linnaeus: *Fauna Svecica* - 2nd edition

[33] 1756 – Linnaeus: *Systema Naturae* - 9th edition

[34] 1767 – Salerne: *L'Histoire Naturelle eclaircie dan une des ses partie principales,*

L'Ornithologie qui traite des oiseaux de terre, de mer et de riviere, tant de nos climats que des pays étrangers.

35 1901 – Leray: "De l'Instinct des Oiseaux dans la Construction de leurs Nids" in *Le Cosmos, Revue des Sciences et de leurs Applications* - Volume 45

36 1902 – Arrigoni degli Oddi: *Atlante ornitologico. Uccelli europei con notizie d'indole generale e particolare*

37 1775 – Buffon / Montbeillard: *Histoire Naturelle des Oiseaux* - tome VI

38 1740 – Frisch: *Vorstellung der Vogel Deutschlandes*

39 1800 – Buffon: *Histoire Naturelle* - volume XLVI or X - Sonnini edition

40 1978 – Simms: *British Thrushes*

41 1784 – Buffon / Otto: *Naturgeschichte der Vogel* - volume 9

42 1795 – Goeze: *Europaische Fauna oder Naturgeschichte der Europaischen Thiere* - vol. V

43 1834 – Baudruillart: *Dictionnaire Général, Historique et Raisonné des Chasses*

44 1823 – Bonnaterre / Vieillot: *Tableau Encyclopédique et Méthodique des Trois Règnes de la Nature - Ornithologie* - volume III & planches

45 1814 – de Bray: *Mémoire sur la Livonie*

46 1826 – Drapiez: *Dictionnaire Classique d'Histoire Naturelle*

47 1836 – *Secrets, Anciens et Modernes de la Chasse aux Oiseaux*

48 1868 – Guardia: Review of "Le Victorial, Chronique de don Pedro Niño" in *Revue Critique d'Histoire et de Littérature* - volume 3, no. 5

49 1824 – Dumont: *Dictionnaire des Sciences Naturelles* - volume XXX

50 1920 – Stresemann: *Avifauna Macedonica*

51 1797 – Bewick: *A History of British Birds* - volume I - 1st edition

52 1776 – Pennant: *British Zoology* - volume 1 - 4th edition.

53 1775 – Ward: *A Modern System of Natural History*

54 1793-1801 – Adelung: *Auszug aus dem grammatisch-kritischen Wörterbuche der Hochdeutschen Mundart*

55 1793 – Nemnich: *Wörterbücher der Naturgeschichte*

56 1796 – Nemnich: *Allgemeines Polyglotten-Lexicon der Naturgeschichte* - volume V

57 1795 – Bechstein JM: *Gemeinnutzige Naturgeschichte Deutschlands* - vol IV - 1st ed

58 1797 – Bechstein: *Grundliche Anweisung alle Arten von Vogeln*

59 1807 – Bechstein: *Gemeinnutzige Naturgeschichte Deutschlands* - vol III - 2nd ed

60 1772 – Dietzschin, Wirsing, Vogel: *Sammlung Meistens Deutscher Vogel* - vol. I

61 1838 – Bechstein: *The Natural History of Cage Birds*

62 1868 – Darwin: *The Variation of Animal and Plants under Domestication*

63 1789 – Naumann: *Der Vogelsteller*

64 1790 – Stübner: *Denkwürdigkeiten des Fürstenthums Blankenburg und des demselben inkorporirten Stistsamts Walkenried*

65 1797 – Naumann: *Naturgeschichte der Land- und Wasser-Vögel* - volume I, folder 4

66 1804 – Naumann: *Naturgeschichte der Land- und Wasser-Vögel* - 1st Nachtrag

67 1822 – Naumann: *Naturgeschichte der Vögel Deutschlands* - volume II

68 1798 – Bertuch: *Bilderbuch fur Kinder* - volume III

69 1798 – Funke: *Ausführlicher Text zu Bertuchs Bilderbuche für Kinder*

70 1807 – Bertuch: *Novus Orbis Pictus Juventuti* - volume V

[71] 1799 – Muller: *Die Vorzüglichsten Sing-Vögel Teutschlands mit ihren Nestern und Eÿern nach der Natur Abgebildet*

[72] 1797 – Bechstein: *Getreue Abbildungen Naturhistorischer Gegenstaende in hinsicht auf Bechsteins Kurzgefasste Gemeinnützige Naturgeschichte des In- und Auslandes*

[73] 1798 – Heppe, JC: *Der Vogelfang* - volume I

[74] 1800 – Bechstein: *Naturgeschichte der Stubenthiere. Vögel.*

[75] 1802 – Bechstein: *Handbuch der Jagdwissenschaft*

[76] 1802 – Hopfner / Gmelin: *Deutsche Encyclopaedie oder Allgemeines Real-Worterbuch aller Kunste und Wissenschaften* - volume XXII

[77] 1803 – Bechstein: *Ornithologisches Taschenbuch*

[78] 1804 – Bechstein: *Getreue Abbildungen Naturhistorischer Gegenstaende* – volume 6

[79] 1804 – Lippold & Funke: *Neues Natur- und Kuntslexicon enthaltend die wichtigsten und gemeinnützigsten Gegenstände aus der Naturgeschichte, Naturlehre, Chemie und Technologie*

[80] 1806 – Orphal: *Der Jägerschule*

[81] 1808 – von Bose: *Neues Allgemein Praktisches Worterbuch der Forst- und Jagdwissenschaf nebst Fischeren* - volume I

[82] 1809 – Campe: *Worterbuch der Deutschen Sprache* - volume 3

[83] 1810 – Albrecht: *Der Kleine Vogelfänger: Eine Hülfsbuch für Jäger, Oekonomen und Vogelliebhaber*

[84] 1812 – Bechstein: *Naturgeschichte der Stubenthiere* – vol. I: *Die Stubenvogel* - 3rd ed

[85] 1816 – Oken: *Lehrbuch der Zoologie*

[86] 1820 – Bechstein: *Die Forst- und Jagdwissenschaft nach allen ihren Theilen für angehende und ausübende Forstmänner und Jäger* - volume I

[87] 1820 – Heinsius: *Volkthümliches Wörterbuch der Deutschen Sprache* - volume 3

[88] 1822 – Winckell: *Handbuch fur Jager, Jagdberechtigte und Jagdliebhaber* - volume III - 2nd edition

[89] 1823 – Brehm CL: *Lehrbuch der Naturgeschichte aller Europaischen Vogel*

[90] 1832 – Nowak: "Die Ringdrossel" *Oekonomische Neuigkeiten und Verhandlungen*

[91] 1834 – Pierer: *Encyclopädisches Wörterbuch der Wissenschaften, Künste und Gewerbe bearbeitet von mehreren Gelehrten* - volume 22

[92] 1834 – Gloger: *Vollständiges Handbuch der Naturgeschichte der Vögel Europa's mit besonderer Rücksicht auf Deutschland*

[93] 1834 – Kaltschmidt: *Kurzgefasstes vollständiges stamm- und sinnverwandtschaftliches Gesammt-Wörterbuch der Deutsch Sprache*

[94] 1834 – Landbeck: *Systematische Aufzählung der Vögel Würtembergs*

[95] 1834 – Hoffmann: *Deutschland und seine Bewohnen* - volume I

[96] 1834 – Fechner: *Das Hauslexicon*

[97] 1835 – Voigt: *Lehrbuch der Zoologie*

[98] 1836 – Grafe & Naumann: *Handbuch der Naturgeschichte der drei Reich für Schule und Haus*

[99] 1837 – Schilling: *Ausfürliche Naturgeschichte des Thier-, Pflanzen- und Mineralreichs*

[100] 1837 – Oken: *Allgemeine Naturgeschichte für alle Stande* - volume VII, section 1

[101] 1837 – Zander: *Naturgeschichte der Vogel Mecklenburgs*

[102] 1838 – Weber: *Allgemeines Deutsches Terminologisches Ökonomisches Lexicon und Idioticon*

[103] 1839 – *Oesterreichisches Naturhistorisches Bilder Conversations-Lexicon*

[104] 1840 – Behlen: *Real- und Verbal-Lexicon der Forst und Jagdkunde mit ihren Hülfswissenschaften*

[105] 1840 – Krünitz: *Oeconomisch-Technologische Encyclopàdie*

[106] 1844 – Train: *Die Nieder-Jagd in allen ihren Verzweigungen zu Holz, Feld und Wasser*

[107] 1845 – Schulz: *Fauna Marchica*

[108] 1846 – Eiselt: *Der Johannesbader Sprudel*

[108a] 1848 – Sandmeier: *Metodisch-Praktische Anleitung zur Ertheilung Eines Geist- und Gemüthbildenben Unterrichts der Naturkunde in Volkschulen*

[109] 1849 – Heyse: *Handwörterbuch der Deutsche Sprache*

[110] 1849 – Heck: *Bilder-Atlas zum Conversations-Lexicon* - volume I

[111] 1849 – Friderich: *Naturgeschichte aller deutschen Zimmer-, Haus- und Jagdvögel*

[111a] 1850 – Sandmeier: *Lehrbuch der Naturkunde, methodisch behandelt für die verschiedenen Stufen der Volksschule* – volume 1

[112] 1852 – Palliardi: *Systematische Uebersicht der Vögel Böhmens*

[113] 1853 – Jäckel: "Zu dem Verzeichniss der Trivialnamen der bayerischen Vögel" in *Naumannia*

[114] 1853 – Speerschneider: "Vergleichende Aufzählung der auf dem Süd-Ost Thüringer Walde und der in der Umgegend von Schlotheim im Nord-West Thüringen vorkommenden Vögel" in *Naumannia*

[115] 1854 – Willibald: *Die Nester und Eier*

[116] 1858 – Tschudi / Winckell: *Handbuch fur Jager, Jagdberechtigte und Jagdliebhaber* - Volume II - 3rd edition

[117] 1860 – Wolf: *Deutsch-Slovenisches Wörterbuch*

[118] 1860 – Grässner: *Die Vögel Deutschlands und ihre Eier*

[119] 1860 – Büchele: *Die Wirbelthiere der Memminger Gegend. Ein Beitrag zur Bayernischen Fauna*

[120] 1861 – Mohn: "Ueber den Unterschied zwischen Schwarz- und Ringdrosseln" *Die Natur*

[121] 1865 – Múller: *Charakterzeichnungen der Vorzüglichsten Deutschen Singvögel*

[122] 1865 – de la Fontaine: *Faune du Pays de Luxembourg*

[123] 1865 – von Homeyer: "Deutsche Singvögel nach A. und C. Müller" *Magazin für die Literatur des Auslandes*

[124] 1866 – Brehm, AE: *Illustrirtes Thierleben* - volume III

[125] 1868 – Sonnenberg: *Die Waidmannssprache: Ein vademecum für Jäger und Jagdliebhaber*

[126] 1868 – Lucas: *Englisch-Deutsches und Deutsch-Englisches Wörterbuch*

[127] 1869 – Opel: *Lehrbuch der Forstlichen Zoologie*

[128] 1869 – Schem: *Deutsch-Amerikanisches Conversations-Lexicon*

[129] 1870 – Berge / Bechstein: *Naturgeschichte der Hof- und Stubenvögel*

[130] 1873 – Russ: *Handbuch für Vogelliebhaber, -Züchter, und -Händler*

[131] 1875 – Meyer: *Konversations-Lexicon* - 3rd edition - volume V

[132] 1876 – Friderich: *Vollstandige Naturgeschichte der Deutschen Zimmer-, Haus- und Jagdvogel.*

[133] 1877 – Giebel: *Giebel's Vogelschutzbuch. Die Nützlichen Vögel unserer Aecker, Wiesen, Gärten und Wälder* - 4th improved edition

[134] 1879 – *Beschreibung des Oberamts Tuttlingen*

[135] 1879 – Brehm / Schmidtlein: *Brehms Tierleben, Kleine Ausgabe für Volk und Schule* - 2nd edition - volume II

[136] 1882 – Reisenthal: *Jagd-Lexicon*

[136a] 1882 – Fraas et al: *Hohentweil. Beschreibung und Geschichte.*

[137] 1883 – Berghaus: *Der Sprachschatz des Sassen* - volume 2

[138] 1883 – Taubman in: *Jahresbericht (1883) des Comité's für ornithologische Beobachtungsstationen in Oesterreich und Ungarn*

[139] 1888 – *Thiel's Landwirthschaftliches Konversations-Lexicon*

[140] 1889 – Koller: "Ornithologische Beobachtungen in Oberösterreich" in *Des Deutschen Vereins zum Schutze der Vogelwelt*

[141] 1891 – Dombrowski: *Allgemeine Encyklopaedie der gesammten Forst- und Jagdwissenschaften* - volume VI

[142] 1890 – Liebe: "Die Gilbdrossel" in *Zeitschrift für Ornithologie und Praktische Geflügelzucht*

[143] 1891 – Bos: *Thierische Schadlinge und Nutzlinge*

[144] 1891 – Jäckel: *Systematische Ubersicht der Vogel Bayern*

[145] 1891 – Friderich: *Naturgeschichte der Deutschen Vögel einschlietzlich der sämtlichen Vogelarten Mittel-Europas*

[146] 1891 – Ferrant: "Ornithologische Notizen" *Bulletin de la Societe Nationale Luxembourg*

[147] 1891 – Olphe-Galliard: *Contributions a la Faune Ornithologique de l'Europe Occidentale* - fascicle 27

[148] 1892 – Brehm / Pechuel-Loesche: *Brehm's Thierleben* - 3rd edition - volume II

[149] 1892 – Floericke: *Versuch einer Avifauna der Provinz Schlesien*

[150] 1892 – Russ: *Handbuch fur Vogelliebhaber, -Buchter und -Handler* - volume II

[151] 1892 – Winteler: *Naturlaute und Sprache*

[152] 1893 – Brehm / Schmidtlein: *Brehms Tierleben, Kleine Ausgabe für Volk und Schule* - volume II

[153] 1894 – Dalla Torre: "Die volkstümlichen Thiernamen in Tirol und Vorarlberg" in *Beitrage zur Anthropologie, Ethnologie und Urgeschichte von Tirol*

[154] 1895 – Löwis: *Unsere Baltischen Singvögel*

[155] 1894 – Meyers: *Meyers Konversations-Lexicon* - 5th edition - volume 5

[156] 1900 – Rey: *Die Eier der Vögel Mitteleuropas*

[157] 1902 – Krichler: *Katechismus für Jäger und Jagdfreunde*

[158] 1903 – Herman: *Nutzen und Schaden der Vögel*

[159] 1903 – Madariaga: *Diccionario Científico-Forestal Aleman-Español*

[160] 1904 – Ribbeck: "Trivialnamen Deutscher Vögel" *Mitteilungen über die Vogelwelt*

[161] 1904 – Bade: *Die Mitteleuropaischen Vogel* - volume I

[162] 1906 – Knauer: "Zur Amselfrage" *Die Umschau*

[163] 1907 – Floericke: *Deutsches Vogelbuch*

[164] 1907 – Schäff: *Jagdtierkunde*

[165] 1908 – Blanchard: *Glossaire Allemand-Francais des Termes d'Anatomie et de Zoologie*

[166] 1909 – Boranić: "Onomatopejske riječi za životinje u slavenskim jezicima" in *Rad Jugoslavenske Akademije Znanosti i Umjetnosti*

[167] 1909 – Herzer: *Českonêmecky slovník*

[168] 1911 – Fatio & Studer: *Catalogue des Oiseaux de la Suisse* - volume 7

[169] 1913 – Brehm AE / Straffen: *Brehms Tierleben* - volume IV

[170] 1915 – "Sprachsaal" in *Die Gefiderte Welt*, pg 23 & 64; volume 44.

[171] 1916 – Heyder: "Ornis Saxonica. Ein Beitrag zur Kenntnis der Vogelwelt des Königsreichs Sachsen" in *Journal für Ornithologie*

[172] 1918 – Tubeuf: *Naturwissenschaftliche Zeitschrift für Forst-und Landwirtschaft* - vol 16

[173] 1930 – Sommerfeld: "Gefiederstúdien an Drosseln" *Anzeigers Ornithologische Gesellschaft in Bayern*

[174] 1941 – Irlweck: *Waidmännisches Lehrbuch für alle Jägerprüfungen*

[175] 1943 – Morbach: *Vogel der Heimat* - volume III

[176] 1957 – Grimm J & Grimm W: *Deutsches Wörterbuch* - volume 19

[177] 1959 – Dornseiff: *Der Deutsche Wortschatz nach Sachgruppen*

[178] 1988 – Cramp: *Handbook of the Birds of Europe, the Middle East and Africa* - vol. 5

[179] 1998 – Desfayes: *A Thesaurus of Bird Names. Etymology of European Lexis through Paradigms*

[180] 2014 – Bertau: *Die Bedeutung historischer Vogelnamen - Singvögel*

[181] 1831 – Hilpert: *A-Dictionary of the English and German Languages*

[182] 1828 – Brehm CL: "Uebersicht der deutschen Vogelarten nach Brehm" *Isis*

[183] 1831 – Brehm CL: *Handbuch der Naturgeschichte aller Vogel Deutschlands*

[184] 1832 – Brehm CL: *Handbuch fur den Liebhaber der Stuben-, Haus- und aller der Zahmung werthen Vogel*

[185] 1839 – MacGillivray: *A History of British Birds*

[186] 1856 – Heinrich: *Mährens und k. k. Schlesiens Fische, Reptilien und Vögel*

[187] 1845 – Brehm: "Einige naturgeschichtliche Bemerkungen auf einer Reise an den Rhein im September und October 1842" in *Isis*

[188] 1855 – Brehm CL: *Der Vollständige Vogelfang. Eine Gründliche Anleitung, Alle Europäischen Vögel*

[189] 1858 – Perini: "Degli Uccelli Veronesi" in *Memorie dell'Accademia d'Agricoltura, Commercio ed Arti di Verona*

[190] 1872 – Salvadori: *Fauna d'Italia*

[191] 1903 – Pavesi: "E sempre il Merlo bianco" in *Rendiconti*

[192] 1867 – Degland & Gerbe: *Ornithologie Européenne* - 2nd edition

[193] 1823-30 – Vieillot: *Ornithologie Francaise*

[194] 1904 – Arrigoni degli Oddi: *Manuale di Ornitologia Italiana*

[195] 1899 – Greene: *British Birds for Cages and Aviaries*

[196] 1909 – Suolahti: *Die Deutschen Vogelnamen*

[197] 1988 – Snow: *A Study of Blackbirds* - 2nd edition

[198] 1919 – Stresemann: "Beiträge zur Kenntnis der Gefiederwandlungen der Vögel" in *Verhandlungen der Ornithologischen Gesellschaft Bayern*.

[199] 1980 – Rohwer et al: "Delayed maturation in passerine plumages and the deceptive acquisition of resources." in *American Naturalist* 115: 400-437

[200] 1985 – Editorial: "Plumage, Age and Moult Terminology" in *British Birds*; 78(9):419-427.

[201] 1997 – Escalona-Segura & Peterson: "Variable plumage ontogeny in the Black (Turdus infuscatus) and Glossy-black robins (T. serranus)" in *The Wilson Bulletin*; 109: 182-4.

[202] 2004 – Gregoire: "Stabilizing natural selection on the early expression of a secondary sexual trait in a passerine bird" in *J. Evol. Biol.* 17:1152-6.

[203] 350 BC – Aristotle: *The History of Animals*

[204] 2016 – Guerrieri: *The Stockamsel: An Historical Review.*

[205] 1802 – Levaillant: *Histoire Naturelle des Oiseaux d'Afrique.*

ILLUSTRATION CREDITS

Front cover: Photograph © Anthony McGeehan. With kind permission of
   Anthony McGeehan.
Page 3: Naumann, *Naturgeschichte der Vögel Mitteleuropas* - volume I (1905).
   [http://www.biodiversitylibrary.org/item/107540]
Page 7: Photograph © Anthony McGeehan. With kind permission of Anthony
   McGeehan.
Page 14: Döbel, HW, *Jäger-Practica* (1746). [http://data.onb.ac.at/rec/AC09816546]
Page 16: Hubertusburg (colored copper engraving by G. F. Riedel, circa 1763).
   Photograph by Foto H.-P. Haack. Wikipedia Commons. Public domain.
Page 18-21: Kramer, *Elenchus Vegetabilium et Animalium Per austriam Inferiorem
   Observatorum* (1756). [http://resolver.sub.uni-goettingen.de]
Page 23 (top): Willughby, *Ornithologiae* - volume I (1676). [http://biodiversity
   heritagelibrary.org/item/129443]
Page 23 (bottom): Aldrovandi, *Ornithologiae - Tomus Alter* - liber XVI (1600)
Page 24 (top): Spotted Nutcracker. Wikipedia Commons. [Released into the
   public domain for any purpose without conditions by Murray B.
   Henson at en.wikipedia]
Page 24 (bottom): Ray, *Synopsis Methodica Avium & Piscium* (1713).
   [http://biodiversityheritagelibrary.org/item/127717]
Page 29: Frisch, *Vorstellung der Vogel Deutschlandes* (1740). [www.staats
   bibliothek-berlin.de]
Page 34: Pennant, *British Zoology* - volume 1 (1776).
Page 36: Bechstein, *Gemeinnützige Naturgeschichte Deutschlands nach allen drey
   Reichen* (1795). [http://www.biodiversitylibrary.org/item/95700]
Page 40: Dietzschin, Wirsing, Vogel, *Sammlung Meistens Deutscher Vogel* (1772)
Page 44: Naumann, *Der Vogelsteller* (1789). [http://www.mdz-nbn-resolving.de]
Page 46: Naumann, *Naturgeschichte der Land- und Wasser-Vögel* (1797).
   [http://ngcs.staatsbibliothek-berlin.de]
Page 48: Naumann, *Naturgeschichte der Land- und Wasser-vögel des nördlichen
   Deutschlands und angränzender Länder* (1797)
Page 50: Bertuch, *Bilderbuch fur Kinder* (1798)
Page 55: Brehm CL, "Uebersicht der deutschen Vogelarten nach Brehm" in *Isis*
   (1828)
Page 56: Brehm CL, *Handbuch der Naturgeschichte aller Vogel Deutschlands* (1831)
Page 57-58: Photographs courtesy of Paul R. Sweet, Collection Manager,
   Department of Ornithology, American Museum of Natural History, NYC.
Page 60: Photograph courtesy of Matthew Shanley, Staff Photographer, American
   Museum of Natural History, NYC.
Page 62: "Christian Ludwig Brehm, ornithologist" by Carl Werner (ca. mid 19th
   century). With kind permission. © Victoria and Albert Museum, London.
Page 65: Brehm CL, *Der Vollständige Vogelfang* (1855)

Page 72: Múller, *Charakterzeichnungen der Vorzüglichsten Deutschen Singvögel* (1865). [http://www.biodiversitylibrary.org/item/29075]

Page 73: Muller, *Wohnungen, Leben und Eigentümlichkeiten in der höheren Thierwelt* (1869)

Page 84: Sommerfeld, "Gefiederstúdien an Drosseln" *Anzeigers Ornithologische Gesellschaft in Bayern* (1930); with kind permission of Manfred Siering.

Page 87: Cramp, *Handbook of the Birds of Europe, the Middle East and Africa* - vol. 5 "Tyrant Flycatchers to Thrushes" (1988). By permission of Oxford University Press (www.oup.com)

Page 88, 90, 92: Photographs © Anthony McGeehan. With kind permission of Anthony McGeehan.

Page 97: Design and photographs © Anthony McGeehan. With kind permission of Anthony McGeehan.

TEXT CREDITS

Page 1-2: © Keith Vinicombe, Alan Harris and Laurel Tucker, 2014, 'The Helm Guide to Bird Identification', Christopher Helm, used by kind permission of Bloomsbury Publishing Plc.

Page 4-5: Snow: *A Study of Blackbirds* - 2nd edition (1988) © D. W. Snow.

[All passages originally written in German, French Italian or Latin have been translated into English].

ACKNOWLEDGEMENTS

I wish to acknowledge the help of my mother, Norma, in translating some of the rather antiquated German for this book.

I also greatly appreciate the help of bird photographer and writer Anthony McGeehan who kindly allowed me to include his photographs of a 'stockamsel' type of Blackbird. Anthony is also author/co-author of several books in which birds come alive with his expressive and entertaining writing: *Birds: Through Irish Eyes* (2012), *Birds of the Homeplace: The Lives of Ireland's Familiar Birds* (2014), and *To the Ends of the Earth: Ireland's Place in Bird Migration* (2018).

Many thanks go to Paul R. Sweet, Collection Manager, Department of Ornithology, American Museum of Natural History; as well as Matthew Shanley, Staff Photographer, American Museum of Natural History, for graciously providing images of Brehm's *Merula truncorum*.

INDEX

About the Author

Claudio Guerrieri is author of *The Stockamsel: An Historical Review* (2016), an in-depth study of the history of the Stockamsel. This hardcover book is recommended for those who wish to delve deeper into the lost history of the Stockamsel and enjoy the original texts written in various languages that chronicle the birth, controversies, and demise of this cryptic bird.

Having resided during his youth in the English countryside and reared on the *Observer's Book of Birds* and of *Birds' Eggs*, it is of little wonder that one of his amateur passions has always been the world of ornithology. He has also written the award-winning *The John Fahey Handbook* - volumes 1 & 2 (2013-14) and *"De Intestini Duodeni Situ et Nexu"* of *Laurentius Claussen* (2018).

He was born in London, England, and graduated from the University of Rome "La Sapienza" with a degree in Medicine. His profession includes a brief period as Medical Lieutenant in the Italian Armed Forces and working as an Anatomic Pathologist for 30 years. He has published research papers principally in the field of gynecologic pathology. For the past 20 years, his secondary focus has been representational oil painting.

Claudio Guerrieri has lived in England, Italy, USA, the Netherlands, and Sweden. He currently lives and works in New Jersey with his wife Aly and daughter Lucia.

www.ingramcontent.com/pod-product-compliance
Lightning Source LLC
Chambersburg PA
CBHW072146020426
42334CB00018B/1897